.

Participatory Democracy, Science and Technology

Also by Karl Rogers

ON THE METAPHYSICS OF EXPERIMENTAL PHYSICS

MODERN SCIENCE AND THE CAPRICIOUSNESS OF NATURE

Participatory Democracy, Science and Technology

An Exploration in the Philosophy of Science

Karl Rogers

palgrave
macmillan

First published 2008 by
PALGRAVE MACMILLAN
Houndmills, Basingstoke, Hampshire RG21 6XS and
175 Fifth Avenue, New York, N.Y. 10010
Companies and representatives throughout the world

PALGRAVE MACMILLAN is the global academic imprint of the Palgrave Macmillan division of St. Martin's Press, LLC and of Palgrave Macmillan Ltd. Macmillan® is a registered trademark in the United States, United Kingdom and other countries. Palgrave is a registered trademark in the European Union and other countries.

ISBN 13: 978-0-230-52206-0   hardback

This book is printed on paper suitable for recycling and made from fully managed and sustained forest sources. Logging, pulping and manufacturing processes are expected to conform to the environmental regulations of the country of origin.

A catalogue record for this book is available from the British Library.

A catalogue record for this book is available from the Library of Congress.

10   9   8   7   6   5   4   3   2   1
17   16   15   14   13   12   11   10   09   08

Transferrred to Digital Printing 2012

For Ryan Johnson-McCourt

"To truly succeed as an emancipatory force, for the free initiative of all and everyone, the revolution must develop freely in a thousand different ways, corresponding to the thousand different moral and material conditions in which people now find themselves. And we must put forward and carry out as far as we can those ways of life that best correspond to our ideals. But above all we must make a special effort to awaken in the mass of people a spirit of initiative and the habit of doing things for themselves."

Errico Malatesta, "Question of Tactics", 1931

"History teaches us that men behave wisely when they have exhausted all other alternatives."

Billy Connolly, *Still Crazy*, 1999

Contents

1
The Call for Democratic Participation

In this era of "democracy promotion" we need to ask tough questions about the nature of the kind of "democracy" that is being "promoted" throughout the world. What do we mean by "democracy"? What does it mean to live in a "democracy"? Do we live in a "democracy"? It is essential that we look at this vague and malleable concept, give it some content, and critically examine it. This is particularly important during a time when the method of "democracy promotion" has taken the form of military strategies of "regime change", undertaken in the name of "preserving our way of life" or "our freedoms" by the governments of the United States of America and the United Kingdom of Great Britain and Northern Ireland, ignoring massive popular opposition among the citizenry and objections from around the world. Appeals to basic human rights and international law have been dismissed or ignored, while appeals to "national-interest" and "national-security" have become the standard tropes used in news sound bites and slogans to justify further governmental authoritarianism, secrecy, and aggression. Cynics among us may be right to point out that the current military adventures in the name of "democracy promotion" have nothing to do with promoting democracy, but are the means to gain access to oil and other natural resources, given that the rhetoric of "democracy promotion" and "regime change" seems to be projected towards those countries that (perhaps, coincidentally) are major suppliers of oil or some other natural resource. However, even if we do not share this cynical perspective, we still need to ask why "democracy promotion" does not begin at home, given that in both the United States and the United Kingdom basic civil rights and legal protections are being eroded in the name of "national security", despite popular protests and legal objections. Controversies regarding electronic surveillance, extraordinary rendition, biometric identity cards, secret courts, and even cases of electoral fraud in national elections have been quickly forgotten by the television and press in favour of the rhetoric of "the war on terror". Yet, if we look beyond the affirmations of "our freedoms and values" fed to us by mass media, it becomes questionable whether we live in a democracy at all, especially given that it tends to be these freedoms and values that are being eroded in order to protect "our freedoms and values". Arguably, the purpose of global "democracy promotion" may well be to convince British and American citizens that they live within democracies, as well as to justify piracy and plunder abroad. Arguably, this is nothing new. The historical creation and development of "actually existing democracy" has largely been the outcome of the economic struggles between land owners, urban bourgeoisie, and workers during the transition from agrarian feudal societies to industrial market economies.[1] The combination of rapid growth in industrial capitalism, the influence of international investment on the development of modern nation-states, and the growth of the middle class and industrial proletariat were the dominant factors in the construction of institutional arrangements to mobilise, direct, and constrain popular movements for democratic participation, in order to preserve and expand the privileges, wealth, and power of the economic elite during the deconstruction of feudalism. Arguably, the current vogue for "democracy promotion" around the world is a continuation of the process of globalising the political and economic conditions for corporatism and the further satisfaction of the interests of capitalists. Of course, the development of "democracy" is not only based on the struggles between economic classes. While economic relations and conditions are clearly major factors in shaping institutional arrangements, it is also the case that cultural traditions, family structures, religion, ethnicity, and gender have profound influences on the construction of the nation-state. The historical development and differentiation of "democracy" is contingent upon the historical and cultural development and differentiation of the society from which it emerges as an institutionalised form of governance and organisation. It is an error (or an act of imperialism) to consider "democracy promotion" to have either a universal emancipatory essence or that the exportation and imposition of specific liberal institutional arrangements provide a deterministic conditions for political freedom. Principles for the democratisation of the State cannot be imposed, without seriously undermining its legitimacy. The democratic ideal of "popular sovereignty" is transformed by the processes involved in giving that ideal concrete expression during the historical development and differentiation of theory and practice in different countries and cultures. These processes must come from the civilian population, in their own terms, rather than be imposed by an occupying force, be it either a "liberating" or "revolutionary" army. In this sense, the social being of democracy – the *demos* – should be an ontological emergence of experiences, efforts, goals, strategies, and aspirations, by the people, for the people, throughout a cultural and historical process of internal interactions, conflicts, accommodations, alliances, reforms, and revolutions. In this respect, the emergence of "democracy" must come from within a society, as a social evolution, and if a foreign power attempts to impose "democracy" and "regime change" through military action, it will inevitably distort the historical development and differentiation of that society. Such a violent act of distortion cannot provide the conditions of political freedom at all, but, rather, merely imposes a simulacrum of liberal democracy, which, due to the means of imposition, is neither liberal nor democratic.

However, this book is not a manifesto. It does not offer a detailed critique of "actually existing democracy"; nor does it offer a detailed model of participatory democracy. This book is a response to the growing call for the democratisation of science and technology. What does this mean? How could science and technology be democratised? Do we have a free relation with science and technology? Or have they an internal logic or essence? The argument of this book focuses on a critical and theoretical exploration of the relations between public participation, science, and technology, in order to philosophically examine the possibilities and limits for democratic participation in a modern society. This is of particular importance when we tend to consider the technical realm of science and technology to be essential for societal development, but something that is outside the political realm or, at least, something that should be. Examining of the meaning of "democratic participation" in relation to science and technology involves more than the question of the jurisdictional limits to the imposition of public moral standards and norms, as expressed in laws and regulations, but involves the deeper public deliberation upon the nature of science and technology, especially in relation to the human capacity to change the material conditions of our existence and the horizon of our possibilities. It is essential that the nature of science and technology should be open to ongoing and public deliberation, allowing people to develop their capacity to choose between alternative paths of societal development. The public examination of the meaning of "democracy" should not be understood as a search for common semantic definition, but rather a call for a deeper ontological understanding of the potentiality and limitations of the *demos* in a modern society, allowing people to better fulfil their potential as democratic citizens by coming to their own understanding about what is involved in being a democratic citizen. It is a public examination of the substantive and transformative effects of democratic participation in relation to the public discovery of shared visions of the ideal society, the good life, and human well-being. It is a call for a public, critical examination of human potential, raising fundamental existential questions about the possibility of human rationality, freedom, and equality in modern society. This involves a critical examination of the implicit visions of the ideal society and the world that underscores our understandings of our potential, alongside the critical reflection upon the implicit conceptions and representations of good human character, civic virtue, and citizenship that are also at play in the construction of the institutional arrangements of civic society. In this book, I shall develop the critical theory of technology proposed by Andrew Feenberg.[2] This theory brings together substantivist and constructivist approaches, as he put it, allowing critical hermeneutic reflections on the meaning of technology, as a substantive historical continuum of development and differentiation, to inform sociological analyses of the norms of technological development in terms of socially negotiated choices, intentions, and expectations, which, in turn, direct technological development in accordance with the kind of society we aim to construct. I will discuss substantive theories of technology in Chapter 2 and Feenberg's theory in detail in Chapter 3. In Chapter 5, I shall develop it to include science as well. Briefly put, my position is that the development and implementation of particular sciences and technologies can reproduce and disseminate particular social ideals, norms, relations, structures, and inequalities, which can radically transform the communities and societies within which they are developed and implemented. It is imperative to foster and develop sciences and technologies that facilitate democratic participation, as well as broaden the public understanding of the nature and purposes of science and technology, to optimise democratic participation in the societal exploration of visions of a rational, egalitarian, and libertarian society, which can guide scientific and technological development by opening the rationality and meaning of scientific and technical criteria to public questioning and deliberation, in accordance with the public understanding of science and technology, as well as societal values, norms, ideals, and goods. By discovering and exploring alternatives, alongside an historical examination of current trajectories, we can critically recover and evaluate our potentiality for the rational development of society. We need to move beyond the consideration of democracy as a particular arrangement of public institutions and, instead, examine it in terms of its ideological and existential meanings for the unfolding ontology of social being.

Participatory democracy, science and technology

How should we understand the relations between participatory democracy, science, and technology? How can democratic participation in the directions and content of science and technology be possible, without distorting or perverting the objectivity or rationality of scientific research and technological innovation? Could democratic participation actually provide deliberative means of improving the rational basis for technical decisions about the trajectories and implementation of science and technology in society? These questions will be critically addressed in the light of a maximal theory of participatory democracy, discussing the potential for and obstacles to the application of such a theory to science and technology. Fundamental philosophical questions about the nature and meaning of democratic participation in a scientific and technological society need to be addressed. We need to question the assumption that the political and technical aspects of science and technology are distinct, where the former deals with public goods, and the latter deals with their technical feasibility. Improving the public understanding of science as a means to better address concerns about the equality of public access to the decision-making process is clearly important to increase the transparency of that process and the accountability of politicians and technocrats. Public access to scientific and technical knowledge is essential for genuine opportunities to choose between (or limit) competing directions of research and implementation. However, the call for democratic participation should not be limited to being a call for public deliberation about whether or how to impose morals or priorities upon the trajectories of scientific research and technological innovation. Nor should it be limited to discussing a series of suggested institutional reforms that promise to increase public participation in the directions of research and development. We need to critically examine how science and technology substantively transform human agency, choices, and experiences, and, as a result, provide conditions and impose constraints on the democratic possibilities for societal development. We need to examine science and technology as agents of social change. The democratisation of science and technology would require free relations with science and technology, but can we develop these relations without undermining the conditions under which science and technology flourish? We need to discover whether the democratisation of science and technology could improve the conditions under which they flourish. We also need to show that a technocratic or authoritarian society actually undermines the conditions under which science and technology flourish. Under what conditions could democratic participation improve the scientific and technological development of society? How can ordinary citizens meaningfully participate in deliberations and decisions involving highly specialised and technical areas of scientific research and technological innovation? In his famous book *Autonomous Technology*, Langdon Winner lamented the way that ordinary citizens in so-called democratic societies have submitted to technical experts' decisions regarding both ends and means – passively subordinating the human lifeworld to a complex of technical systems.[3] Winner argued that we need to carefully consider the possible effects of any proposed technological innovation upon human freedom and social justice, as well as other ethical and political implications, prior to developing and implementing that innovation. He advocated that the goals of large-scale technological enterprises should be carefully and critically examined in advance, but, also, the onus is on each one of us to learn more about technology before we commit ourselves as users. Careful consideration must be made in relation to implicit visions of human well-being and the ideal society. Furthermore, as he argued in *The Whale and the Reactor*, technology is not politically neutral and the political implications for democracy must be considered prior to its technical implementation and development in society.[4] Is the proposed technology likely to benefit the whole of society or only a section of it? Who will be empowered or disempowered by this technology? Winner advocated decentralised democratic participation to dismantle technological systems and reconstruct them in accordance with democratic deliberations and decisions regarding human needs. He called for such careful consideration to involve all citizens and their elected representatives, requiring study and debate into the conditions and contingencies of human choices regarding the trajectories of technological innovation. This involves analysing the substantive impact of proposed technologies on societal relations and institutions, as well as human relations and activities in general. In his later work, Winner has been more explicit about how the participatory democracy movement is a crucial stage in bringing technology back under human control and reflection.[5] He argued that ideological presuppositions which underwrite positivistic political science have tended to suppress discussion of the relation between technology and politics, at most only suggesting reforms to alleviate the symptoms, but leaving the causes of the problem of the technocratic domination of society unquestioned. He also argued that this tends towards reinforcing the *status quo*, positioning those who are enfranchised at a high level of expertise, the technocratic elite, as the political elite in the new chambers of power. He was highly critical of the ideological faith in the beneficence of technological innovation and its slavish adherence to the technocratic imperatives that it has inherited from industrial capitalism. For reasons that will become apparent, I am in broad agreement with Winner's argument. As I shall argue in this book, the democratisation of modern society requires greater democratic participation in science and technology, but this requires an active commitment on the part of the citizenry to democracy. The democratisation of science and technology cannot be achieved without widespread public commitment to take responsibility for the democratisation of society by identifying, confronting, and removing the power of undemocratic and antidemocratic controls over society in general by developing alternatives through local assemblies and free-associations. The power inequalities and preservations of class privileges upon which such controls are grounded in both the ideological and practical organisation of science and technology within society can only be removed through a democratic revolution, which requires a popular commitment to that revolution. If we are committed to the democratisation of society, the theoretical challenge facing us is to understand the nature of a democratic revolution in a scientific and technological society.

The relationship between democratic values and technical expertise is a central concern of this book. Chapters 2 and 3 are primarily concerned with relations between technology, human freedom, and democratic participation. Chapters 4 is concerned with a general discussion of theories of participatory democracy. Chapter 5 applies this discussion to scientific research, and Chapter 6 draws the previous chapters together to provide a critical discussion of the possibilities for and limitations of democratic participation in the development of a rational society. I shall question the commonplace assumption that "technical and scientific matters" should be left in the hands of "technical and scientific experts" by questioning the assumption that this leads to good science and technology. After elaborating substantive and critical theories of science and technology, exploring the political and societal values, meanings, priorities, and ideals that underwrite science and technology and the way that they transform human thought and action, I shall argue that the identification and recovery of internal democratic values within scientific research and technological innovation are essential for the rational development and implementation of science and technology, given that the democratic values, such as tolerance, honesty, fairness, openness, and freedom of conscience (which includes freedom of speech and association, as well as the pursuit of morality, meaning, and truth) are also the intellectual virtues inherent to the scientific tradition, alongside the technical virtues, such as carefulness, attentiveness, patience, and rigorousness. Democratic values include the intellectual virtues required for critical and open participation in the scientific community and its interpretation of the scientific tradition. Serious philosophical and political problems for notions of scientific objectivity arise when the scientific tradition embodies social and political values that suppress or distort these democratic values. For example, consider the historical reinforcement of gender and sexual prejudices via scientific representations of evolutionary and biological processes.[6] These have been used to marginalise women in society and to justify the continued exclusion of women from full participation in the scientific community. Once we consider the democratic values as including intellectual virtues, we can see how these societal prejudices against the participation of women in the scientific community led to a lack of critical examination of how these prejudices were underwriting the methodologies and premises of scientific research, and this lack of critical examination prevented a deeper and rigorous examination of how the scientific tradition was being interpreted by the scientific community. Since women have been included in the scientific community, many of these prejudices have been exposed and contested, which has lead to better science by challenging the rationality and objectivity of these prejudices and the methodologies they underwrote. Of course I am not suggesting that only women can challenge gender stereotypes about women and the inequalities they entail, but, it is more likely that irrational assumptions will remain implicit within a homogeneous community, and the participation of women in the scientific community increases the chance of such gender stereotypes being challenged because it increases the heterogeneity of the scientific community, bringing new perspectives and experiences into play, as well as providing counter examples to the stereotypes. Within an open and free process of inquiry, through the participation of even a single new heterogeneous generation, the effects of plurality and diversity inevitably and rapidly result in criticism of otherwise hidden contradictions and incoherence, often exposing and undermining deep seated prejudices and inconsistencies, which can pass on unchallenged from generation to generation in homogeneous communities.

Scientific and technological activities are directed towards the achievement of social purposes, which presuppose normative and progressive conceptions and representations of science and technology as human goods, implicated in the construction of a better world for human beings. This presupposes a shared vision of social ideals. As Daniel Kleinman argued, this shared vision underwrites a "social contract" between scientists and the public.[7] Scientific work should be subjected to public scrutiny, as a system of exchange of funding for public goods, but, more importantly, there needs to be an ongoing process of public deliberation about the societal goals and public goods towards which science and technology are directed towards satisfying. As I argued in *Modern Science and the Capriciousness of Nature*, the importance of democratic participation becomes paramount once we understand that social purposes presuppose visions of the good life (in the sense of a life that is good for human beings to live) and there are not any universally accepted experts in how to envision and realise the good life.[8] Once we are confronted with a plurality of contested visions of the good life, the ideal society, and human well-being, then how do we choose between them? We cannot turn this question of how to achieve it over to "the technical realm" of scientists and technicians. Scientific knowledge and technical expertise are increasingly specialised and, outside of their narrow area of specialisation, scientists and technicians are as much "lay members" of the public as any other citizen. Scientists and technicians are simply not qualified to know how to develop and implement scientific research and technological innovation for the public good in an open-ended, complex, and changing world. Nor are they qualified to know which public goods should be pursued. All citizens are on a par when considering this question. The question of how to choose between public goods should be answered in "the public realm". But, before we can do this, we need to take a step back and examine the potential for democratic participation in modern society. My argument in this book is that democratic participation in local community development promises to increase civic responsibility and local sustainability, in accordance with local visions of a good community, while also increasing societal experimentalism and creativity, without the need for centralised control or bureaucratic administration. Maximal democratic participation promises to provide society with an evolutionary dynamic, alongside allowing the conservation of local traditions wherever these are valued by local people, which allows local people to determine the conditions of their own existence in accordance with their shared cultural meanings, values, standards, and ideals, when they deem these to be satisfactory. This assumes that local people have sufficient knowledge, experience, skills, imagination, and motivation to decide for themselves how best to live their own lives. As Theodore Roszak put it,

> ... if one believes in the validity of participatory democracy (and what other kind is there?), then it is little more than academic presumption to begin unloading a host of institutional schemes in the abstract... people in the process of changing their homes, neighbourhoods, cities, regions, who are most apt to know best what they need and what works. And if they don't know, they will only learn from their failures. Nothing but responsibility for their own lives makes people grow up and be competent. The resourcefulness of ordinary people whose citizenry instincts have been awakened and put in touch with their community can indeed be amazing. Their experienced judgement always counts for more than the most prestigious expert.[9]

Modern science and technology clearly have social value. They bring benefits by discovering how to remove many of the sources of suffering and death, such as disease or disabilities, by providing antibiotics, pharmaceuticals, medical techniques, and an increased understanding of hygiene. Science and technology have reduced the need for toil and other physical limitations by providing tools, machinery, communications, electricity, heating, refrigeration, transportation, agriculture, industry, and information storage, among other things. However, science and technology have also provided weapons and techniques for imprisonment, surveillance, torture, and propaganda, as well as having provided the means to replace local crafts and trades with mechanised assembly lines, intensified labour, increased control over the workforce, and new sources of pollution and means of plundering the natural world. Just as they have been directed towards dominating, pacifying, and exploiting the natural world, science and technology have also been directed towards aiding one group of human beings to dominate, pacify, and exploit others. The same trajectory of research and development that has empowered some human beings, creating a new horizon of possibilities, has also resulted in a permanent war economy, fuelled by the globalisation of intensified capital accumulation, placing incredible power at the disposal of greed and egoism, which has resulted in an unprecedented level of environmental and social degradation and destruction, at the expense of many other human beings. The same trajectory that has enhanced the human potential for freedom from the limits of our natural state of being also threatens our survival (as well as many other species) and we should not be complacent about the outcome of scientific technological activities. We should not assume that this trajectory is well understood and controlled by scientists, technocrats, professional politicians, or the CEOs of multinational corporations. As I argued in *On the Metaphysics of Experimental Physics*, modern science and technology are aspects of an ongoing societal experiment in the construction of a technological framework of overlapping specialisations and there are not any experts in science at all.[10] At best, a scientist is an expert in a narrow specialisation of science, in the sense of being experienced in the application and refinement of a set of techniques, devices, and theories, and the high degree of uncertainty and ambiguity in a complex, open-ended, changing world actually undermines the claims to expertise made by some scientists and technicians, even in their narrow specialisation. The meaning of any theory or practice can only be understood in hindsight, through their usage in context, requiring interpretations and judgements, while their truth and value are always deferred to the future, subject to testing through ongoing technological activity and innovation, open to revision as theory and practice are refined and modified to accommodate the particularities and problems involved in testing, implementation, and development. Each observation is itself tentative and provisional, requiring skilled interpretations and judgements on how best to make and communicate the observation, and research and development lead to unintended and unpredictable consequences in the wider world outside the laboratory. Such effects transform society in quite unforeseeable ways that cannot be derived from abstract laws because these transformations occur in context, through decisions about usage and purpose, in relation to the particularities of experience of the consequences of our actions. This, in turn, transforms science by creating new possibilities, expectations, and needs, which creates new experiences, experiments, and observations. This process is dialectical, within an historical continuum of development and differentiation, but, as Andrew Pickering argued, every technology is itself a product of a series of accommodations to resistances during its research and development, and, more often than not, this involves a radical departure from the original conception of the purpose of the technology in question.[11] Science and technology are creative social activities that bring new things into the world, allow the realisation of otherwise unachievable intentions, and change how we understand both the world and ourselves. They extend our horizon of possibilities, as well as placing discipline, demands, and expectations on human thought and agency. They embody human values, ideologies, and intentions in their ongoing construction and usage, thereby disseminating and reinforcing these human values, ideologies, and intentions in their methodologies. The implementation and development of scientific research and technological innovations in society involve complicated, diverse, and often incompatible political, economic, and aesthetic goals, choices, demands, and expectations. Science and technology emerge from an evolutionary trajectory, adjusting its operations and scope in relation to the changes it causes in its environment and the problems that arise as it interacts with pre-existing technologies and the natural world. All technical judgements are made in context and technical rationality requires adaptation and accommodations to the particularities of the social organisation and trajectory of any scientific experiment or technical project. Technical rationality is bounded by the substantive meaning and empirical content provided by the real situations that emerge during implementation and development. Neither science nor technology can be separated from its organisation and trajectory within society, without removing all substantive meaning and empirical content from their definition. Conceptions of scientific and technical progress are intimately bound together.[12]

"Actually existing democracy"

"Democracy" has contested meaning, both in theory and practice. David Held argued that the history of "democracy" shows that there are conflicting interpretations of the meaning of "democracy".[13] Not only do these interpretations vary between different people within different cultures, as well as during different historical periods, but the modern usage of the word "democracy", given in terms of the contemporary standards and institutions of "actually existing democracy", tends to be based on the reification of institutional arrangements that have been imposed upon the majority by a wealthy and powerful minority, for the purpose of preserving their political and economic privileges, which are justified, using vague, ambiguous, and inconsistent concepts and representations as being either the best form of government or the least worst, given the nature of human beings and the world. "Actually existing democracy" has the form of an indirect democracy, wherein the citizenry elect their representatives or leaders, the institutional arrangements of the State are supposed to protect individual liberties and property rights, and economic activity is considered to be a private affair. It has the structure of a technocracy wherein governance is left to technical experts appointed by elected politicians, a centralised bureaucratic administration has grown to provide and regulate public services, and the State legislates the development of civic society. This is called "representative democracy" and is very different from direct or "classical" democracy. Advocates of representative or indirect democracy tend to respond to a quite crude notion of direct democracy, which tends to be represented as either centralised "majority rule" or an *ochlocracy* (disorder of the rabble). This ignores many of the alternative kinds of direct democracy, including participatory democracy, and dismisses them out of hand as not being worth the risk to individual liberties and private property. Since Walter Lippman published *The Phantom Public*, advocates of representative democracy have taken it for granted that the majority of people are incapable of governing their own affairs and that the *demos* is a myth of classical literature or philosophy.[14] Lippman advocated representative democracy based on competitive elitism, wherein "majority rule" should be limited to voting for their leaders. Max Weber also advocated competitive elitism as the best method to govern an advanced industrial society.[15] He argued that centralised bureaucratic administration is indispensable to provide a predictable and stable administration capable of dealing with the complex and differentiated problems. He argued that it is inevitable that any direct democracy would construct a centralised bureaucratic state apparatus to unify society and prevent deliberations and decisions from degenerating into a quagmire of infighting factions and civil war. Under such circumstances, technocratic control of society could only be prevented through the democratic election of good leaders. Weber's reliance on the moral integrity of elected leaders was based on the assumption that those capable of gaining power through competitive elections were the best suited to exercise it.[16] This model of "representative democracy" was further developed by Joseph Schumpeter.[17] He argued that the empirical facts about human nature did not support direct democracy or "classical theories of democracy" (such as those proposed by Aristotle, Rousseau, or John Stuart Mill) as being plausible for modern society and, instead, he proposed a more empirically grounded theory of democracy based on "the facts of political life". He argued that an adequate theory of democracy does not need to make reference to any inherent ideals, norms, or virtues, and democracy should not be considered to be an end-in-itself. Democracy should only be evaluated as a method to achieve a stable institutional arrangement between government and the citizenry. According to Schumpeter, once we reject any inherent democratic values, norms, or ideals, the empirical facts of political life, as he saw them, showed that the democratic method should be understood as a competitive election process to select representatives within an institutional arrangement to protect individual rights, administrate societal needs and demands, and protect private property. Elections should operate like a market, within which political parties nominate and support the candidates they deem most likely to "sell" themselves and the party's policies to an electorate of "policy consumers". Democratic participation would be strictly limited to either voting or competing for votes. Discussion between citizens about candidates would be a private matter. Even though anyone could, in principle, compete for the popular vote, Schumpeter considered that those with the inborn capacity for leadership, including intellect and moral character, would naturally comprise a ruling class, from which viable candidates would promote themselves or receive support from a political party. Providing that enough citizens participated in the election of their representatives, the democratic method would work most effectively if elected leaders were left to make decisions without any need to be accountable to the public, until the next election.[18] He even argued that petitioning representatives should be prohibited because it would interfere with effective leadership.

For the advocates of competitive elitism, "democracy" does not mean "rule by the people", but, rather, that people have the opportunity to decide who their rulers are. "Democracy" is "the rule of the politician", as Schumpeter put it.[19] He argued that there is not any common good towards which "all people could agree on or be made to agree on by the force of rational argument" and any society trying to decide "the common good" would inevitably result in supporting and serving the interests of leaders, and, consequently, the only democratic possibility was for the people to choose between competing leaders.[20] Like Lippman and Weber, Schumpeter assumed that those who successfully won elections were (somehow) more knowledgeable, experienced, and competent to govern public affairs than the electorate. Competitive elitism assumes that those who acquire power are best suited to exercise it. This presupposes that the knowledge of the public good is somehow derived from the knowledge of how to become a member of the political elite by gaining public support, as if moral integrity or wisdom were derivative from persuasiveness and popularity. However, even if we accept that this implausible claim is true, it is not clear why it is assumed that an adversarial "winner takes all" competition adequately addresses societal pluralism regarding public goods. One would be tempted to suggest that some kind of system of proportional representation would be more appropriate, especially given that, if there are more than two candidates in an election, more often than not, elected representatives do not have majority support, even when they gained more votes than any one of their competitors. Moreover, Schumpeter claimed that "the popular will" is an irrational social construct that can be manipulated by advertising and propaganda.[21] If we accept this claim, then how could we reasonably expect the outcome of a competitive election to lead to the best choice? It would seem that competitive elitism would favour those best skilled at conducting advertising and propaganda campaigns, rather than those most likely to direct the state apparatus to satisfy the public good for the greatest number of people. It seems that Schumpeter's democratic method tends to place the state apparatus in the hands of those with access to wealth and media who have successfully demonstrated themselves to be most capable of deceiving and manipulating the public. The contradiction at the heart of competitive elitism is that it assumes that the majority of people lack sufficient intellectual or moral standards to govern their own affairs, but, somehow, they possess the ability to intuit who is best qualified to govern from among the available candidates. I term this as "the magical theory of mandate", whereby it is assumed that legitimacy is distinct from the consensus opinion, but consensus opinion is somehow sufficient to legitimate elected leaders. This simply assumes that people can only be trusted to know what they want and are the best placed judges to decide whether any candidate will satisfy those wants. The legitimacy of "representative democracy" paradoxically requires that individual citizens are capable of rationally deciding what their self-interest and preferences are, in order to make an informed decision about which representative is most likely to serve those interests and preferences, but, competitive elitism assumes that the majority of citizens are incapable of making rational decisions. As Carole Pateman argued, "representative democracy" requires rational self-interested individuals, but these are the exception rather than the norm.[22] Citizens are subjected to increasingly subtle and pervasive mass media propaganda campaigns, paid for by pressure groups and political parties, while special-interest groups are able to exert a disproportionate level of influence on the elected representatives by providing the necessary campaign funds to pay the costs of mass media access required for re-election.[23] Once we take mass media and special-interest groups into account, it seems that Schumpeter's democratic method does not fit "the facts of political life" either. Given his obvious contempt for the majority of people, as being too feebleminded and ignorant to be allowed to interfere in political matters, it seems quite hard to understand why he was concerned with the democratic method at all.[24] How could one expect ignorant and feebleminded people to choose competent and honest leaders? After all, how can we expect anyone unfit to govern themselves to be fit to choose who should govern everyone else? Surely it would be better for the majority of people not to be involved in the democratic method at all! It seems that the democratic method has little to do with choosing the best leaders, but, arguably, it is a form of crowd control by pacifying the political urges of the majority and channelling their energies away from political action. It is a method for the legitimation of professional politicians, rather than discovering and deliberating those qualities best suited to statesmanship. It is not clear that competitive elections do, in fact, provide the best method for choosing the best leaders. On Schumpeter's account of the intellectual and moral capabilities of the majority of people, it would be better to choose leaders by lottery.

However, if the electorate are capable of choosing those best suited to realise the public good, then the electorate must also be capable of making good judgements about issues, policies, and the development of society, given that the electorate are supposedly making such judgements when choosing between candidates on the basis of their manifesto or platform. This would undermine the presumption that elected representatives are more capable of making the value judgements at play in deciding questions regarding the goals of societal development. It is not at all clear that there is any logic connection between being able to make practical or rational judgements and being able to win an election. They seem to be independent abilities and require very different skills, knowledge, experience, and strategies. If this is true then competitive elitism is concerned only with legitimating rulers, rather than deciding who is best qualified to rule, and, in the absence of any universal agreement on the qualifications for leadership, Schumpeter's democratic method is little more that a process wherein the largest minority is empowered to impose its choice of leader(s) on all others. This suppresses all alternatives and prevents reasoned deliberation, until the next election campaign, and, by reducing the democratic method to the means to choose between leaders, "the sovereignty of the people" must be suppressed by the State to preserve the democratic method. Even if competitive elections legitimise the rulers by making them accountable, it does not follow from this accountability that elected leaders are either competent or honest. Given that the competitive multi-party system leads to the promotion of expedient strategies and tactics to gain support, rather than necessarily decide the best course of action, competitive elections tend towards an *adhocracy* rather than a representative democracy, while also being vulnerable to the problems of direct democracy, such as factionalism, irrationality, corruption, demagoguery, popularism, and the cynical manipulation of public fears and ignorance. Moreover, as well as being vulnerable to the flaws of direct democracy, the authority and power of elected leaders are secured through the authority and power of the state apparatus and, as a result, it is more difficult for people to directly challenge and change the system, should it become corrupt or despotic, than they could in a direct democracy. Given "the facts of political life", it should be quite unsurprising if a large proportion of the citizenry feel alienated by "actually existing democracy". It is questionable whether public apathy is the result of inherent inaptitude or indifference to politics on the part of most people, as Schumpeter asserted, or whether it is the product of the systematic exclusion of the majority from genuine opportunities for participation. Arguably, "the facts of political life" are little more than Schumpeter's whiggish interpretations of the *status quo*. Alternatively, it may well be the case that public apathy and cynicism are consequences of competitive elitism and public manipulation, rather than good reasons for them.

Schumpeter argued that civil liberties, tolerance of the political opinions of others, and a national allegiance to the "structural principles of existing society" were conditions for the successful operation of the democratic method, but he argued that the democratic method should not be relied on to maintain these conditions.[25] It could as easily result in unjust or immoral decisions, as it could result in just or moral decisions and, therefore, should not be considered as a moral method or end-in-itself. Hence, the democratic method could result in the persecution of a religious, ethnic, or political minority; we should not accept this result simply because it was a result of the democratic method. This opposes direct or "classical" theories of democracy, which tend to treat democratic participation as an end-in-itself and accept whatever the majority decided as being the correct result. However, which moral standards should override the results of the democratic method? Schumpeter appealed to the moral standards of traditional liberalism, wherein political freedom is defined in negative terms as the freedom from coercion and unwarranted outside interference upon either the choices or privacy of the individual. Democratic values, such as freedom of speech and association, necessarily require the protection of the private realm (of individual citizens) from unwarranted interference from the State. The "burden of proof" must rest with the State to justify its claims that its regulation or interventions are warranted and legitimate in order to best protect the equality and rights of all individual citizens. The State's authority to intervene in order to create the conditions for "good citizenship" or the satisfaction of "national interests", as decreed by the agents of the State, must be strictly limited by constitutional law in order to maximise the political freedoms of individual citizens to discover and decide for themselves what constitutes "good citizenship" or "national interests", providing that they respect the constitutional rights of other citizens. In this respect, the liberal tradition of constitutional law is essential to protect the plurality and diversity of a heterogeneous society by providing individual citizens with as much latitude as is possible to simultaneously live in whatever way they consider to be the best possible and also respect the right of others to do likewise. The philosophical problems in the liberal tradition largely are the result of tensions between these two (often conflicting) demands between positive and negative freedom, as Isaiah Berlin famously termed them, where the former is defined as the freedom to act upon one's choices or will, and the latter is defined as the freedom from coercion or oppression.[26]

Berlin's concept of "negative freedom" implies a duty to respect the liberty of others as the basis upon which one expects others to respect one's liberty – a social contract, in Rousseau's sense – providing a socially emergent commitment to norms and limits, without which "positive freedom" cannot be anything other than the individual satisfaction of the will to power. It is only within the socially emergent boundaries of "negative freedom" that "positive freedom" can emerge as an uncoercive form of liberty, where one's "positive freedom" is dependent upon the mutual respect of all for the "positive freedom" of all. Without "negative freedom", "positive freedom" is little more than "the last man standing" outcome of power struggles and conflicts; it is the outcome of biological behaviour, rather than political action (if we take choice and deliberation to be essential aspects of political action). Negative freedom implies a presumption in favour of positive freedom in the sense that the State needs to justify claims for the constraint of particular positive freedoms on the basis that they impose upon the negative freedoms of others. For this reason, liberal pluralists, such as William Galston, have argued that the institutional balance between positive and negative freedom must promote the "maximum feasible accommodation" of "differences among individuals and groups over such matters as the nature of the good life, sources of moral authority, reason versus faith, and the like". [27] The constitutional protection of individual freedoms and rights, in a heterogeneous society, is a minimal condition for the protection of minorities and the preservation of cultural diversity and pluralism, which protects the societal stock of standards, meanings, and aspirations that allow individuals to make meaningful choices. Even if any minority does not place a high value on individual choice and its members discourage each other from exercising it, interventions can only be justified by the State *if and only if* any individual wishes to be freed or dissociated from any minority group and the other members will not permit it. Hence conservative traditions and fundamentalist religions should be respected as equally legitimate as liberal pluralism under liberal constitutional law. The opportunity for exercising positive freedom must be available for all citizens, but they are not under any obligation to exercise it. For example, some religions are extremely prescriptive regarding the public behaviour and appearance of women, while also preserving inequalities between men and women regarding what is considered to be a permissible aspiration and expectation. Many of the prescriptions and inequalities are illiberal and some seem quite incompatible with the liberal tradition. However, unless the agents of the State satisfy the "burden of proof" to provide unequivocal evidence that any particular woman is being directly oppressed by others, by abusing her rights against her will or preventing her from exercising her rights, it must be assumed that she has freely chosen not to exercise particular positive freedoms and she is complicit in her acceptance of these religious prescriptions and inequalities. From the liberal pluralist perspective, the State cannot legitimately intervene based on claims of "repression" or "false consciousness" on behalf of the good of the individual concerned. It is not the responsibility of the agents of the State to decide what is true or good on behalf of the citizens (beyond preserving their individual rights and protections) and the exercise of positive freedoms cannot be imposed without making a mockery of the concept. The State is strictly limited to providing liberal constitutional laws as an overriding legal protection, equally available to all citizens. The imprisonment of women, forced marriage or domestic enslavement, genital mutilations of children, or any such oppressive or violent acts, would be prohibited under liberal laws, but it is acceptable for particular women to consent to traditional practices, say arranged marriage or the use of veils, providing that, at any time, they are able to dissent from this practice, or dissociate themselves from their religion, without fear of persecution. Citizens do not have an obligation to be liberals, and, unless they specifically request legal protection from agents of the State or they are being overtly subjected to violence or oppression, the obligations of the agents of the State are limited to informing citizens of their legal rights. In this respect, tolerance, diversity, and pluralism are maximised at the societal level, while any particular individual citizens can live according to whatever epistemological or moral standards they find most appealing or appropriate for living a good life. In this sense, negative freedom can be understood as the positive freedom of disassociation, whereby liberal pluralism protects the right to dissent and explore alternative ways to discover and realise human goods and ideals. Pluralism is not relativism. It is progressive in the sense that it allows citizens to have opportunities to decide from a cultural stock of competing moral claims (including duties, rules of conduct, and obligations), values and goods, beliefs and aspirations, and commitments to existential projects, while also learning from the experiences of others attempting to live by these standards. The difficult choices are between different goods, rather than between good and bad, and liberal pluralism recognises the absence of any universally accepted standards that would lead us to agree on how we should make such choices. Positive freedom is relegated to the private realm as being a matter for the satisfaction of subjective preferences through traditional practices, free market exchanges, or personal choice. It was for this reason that advocates of competitive elitism and liberal democracy, such as Lippman, Weber, and Schumpeter, have argued that there is a close connection between the democratic method and liberal capitalism, wherein the election of representatives and the "invisible hand" of the free-market would counter the technocratic tendency towards centralised planning, regulation, and control.

However, as I argued in *Modern Science and the Capriciousness of Nature*, capitalism and democracy can conflict; they are not necessarily compatible and congruent.[28] The ideals of liberal capitalism may well be those of a free-market and individual liberty, prosperity, and entrepreneurship, but, once the capacity for the accumulation of capital was empowered by organised science and technology, liberal capitalism was transformed into industrial capitalism.[29] Within liberal economic theory, inequalities in research capabilities and access to supporting technological infrastructure are completely ignored; consequently, the individual entrepreneur or inventor is treated as if they have equal market access as enjoyed by a multinational corporation. It assumes that market interactions are inelastic. However, once we take the scientific and technological inequalities into account, we can address how the resultant leverage and economies of scale select in favour of large businesses, such as multinational corporations, which can raise the costs of market access (advertising costs, for example), as well as take advantage of cheap labour markets, access to governmental contracts, and favourable legislation. Once the means for research and development have become private property, capital accumulation combined with the increasing costs of market access tends towards the domination of the market by the economic elite, which inevitably results in the control over the structures, regulation, and development of the market by powerful oligopolies and monopolies. The private ownership of technological and scientific advances has further centralised and concentrated political power in the hands of the economic elite by allowing them to exercise control over markets, media, and the forces of production, while the State is increasing dependence on the private sector to supply public services. Under such conditions, it is naïve to assume that science and technology have some essential or value-neutral self-correcting mechanism to ensure its objectivity. The economic elite increasingly impose "market discipline" upon the content and trajectories of scientific research and technological innovation. When the economic elite have a monopoly on deciding how science and technology will be developed, the State has increased dependency on the economic elite, and they will have increased influence on deciding how modern society should be developed. As a result, the majority of people find the economic conditions of their existence controlled by a few wealthy individuals, and democratic participation is increasingly limited to consumer choice between corporate supplied services and products. The economic elite determine the conditions for access and exchange within the market, which is only a free market (in the liberal sense) for members of the economic elite to compete with each other for lucrative contracts. The majority of people are compelled to either sell their labour in accordance with the terms set by the market, starve, or resort to criminal action. The so-called free-market is based upon the establishment of power relations created through negotiations between the agents of the State and a minority of owners of the means of research, development, production, and distribution. The accumulation and concentration of capital, alongside the technical rationalisation of large enterprises, have lead to the expansion of the planning of output, prices, and investment by corporations in combination with the growth of centralised state control and regulation to deal with increasing demands, expectations, and conflicts. The State regulation of the relations of production and consumption serve the interests of this powerful minority by protecting their private property and the power relations upon which the market depends. Increasingly, multinational corporations control the means to produce and provide our food, clothing, utilities, transportation, communications, and housing. The basic needs of human existence are increasingly controlled by the economic elite. Multinational corporations provide our entertainment, sports facilities, hospitals, schools, arts, childcare services, health care services, higher education, information storage and retrieval, mass media, and scientific research. The political realm has been in large part reduced to being the means to better integrate the institutional arrangements of the State with the needs of multinational corporations. The infrastructure of society has become the private property of the economic elite, while societal structures (such as laws, institutions, and public administration) are increasingly constrained and directed in accordance with the business strategies of multinational corporations. This greatly reduces the criteria under which societal development is conducted. The technocratic development and rationalisation of the forces and relations of production under industrial capitalism has resulted in the intensified conglomeration of the ownership of the means of production, consolidating corporate control over industries, services, and transportation, where the accumulation of capital is concomitant with the accumulation of the control over scientific and technological development, further increasing the benefits of leverage and the economies of scale, which results in intensifying the inequalities of market access and power relations, and, due to the increased dependency upon campaign funds for the re-election of representatives, there is a selective tendency for elected representatives to favour corporate interests. This will inevitably lead to a kind of corporate-state, a technocratic and totalitarian mass society, which will undermine liberal capitalism and civic society will become absorbed into a system of centralised planning of the total system of production and consumption in terms that favour the economic elite. The liberal tradition is being eroded by the emergence of global corporatism, which is leading to an economic system of planning and control dominated by the economic elite via increased power of multinational corporations over the institutions of nation-states, due to the increased dependency of nation-states upon multinational corporations as producers, suppliers, and service providers.

Of course, economic development is undeniably a condition for liberal democracy, in the sense that a wealthy country is more likely to develop robust and stable liberal institutions and systems of representation than a poor country, even if it is not necessarily the case that a wealthy, industrial country will develop into liberal democracy.[30] However, the contents of the formal features of liberal democracy (such as universal suffrage, elections, freedom of speech and assembly, rule of law, constitutional rights, accountability, transparency, impartial judiciary, and a liberal capitalist market) are substantively contingent upon the historical process of constructing state institutions in interaction with all the other participants in society, as well as dealing with interventions and influences from other countries and cultures. The actual processes of integrating and developing any institutional arrangements give these arrangements their meaning and function, which are not necessarily identical with the intentions and hopes of their "founding fathers". Any failure to pay close attention to this historical and cultural contingency in the development and differentiation of democracy around the world will inevitably distort how that process is understood and, as a result, will lead to distorted theories and practices, possibly amounting to little more than propaganda and colonisation. Once cannot say that any nation is a "democracy" because "it protects freedom of speech", for example, without attending to what "freedom of speech" actually means in the context of everyday practices, how these practices relate to the institutional structures of the State, and how the constitutional legal framework for civic society actually relates to the activities of the agents of the State. It remains possible that while a nation-state may be formally democratic, i.e. hold itself to be democratic due to its history and democratic constitution, in actuality, in practice, it may well be deeply oppressive and undemocratic, in the terms of its own democratic constitution. After all, the political freedom of free speech does not amount to much if it only applies to the speech acts that the agents of the State consider acceptable or permissible. We need to look beyond the formally defined constitutional function of the institutions of the State and look closely at their substantive effects upon the possibilities and limits for genuine democratic participation in exercising political freedom and deciding the directions of societal development. The notion of "genuine democratic participation" must remain an ideal, even as an abstraction, against which the concrete "actually existing democracy" can be judged in terms of whether it facilitates more or less public participation in defining or changing the institutions of the State, in both its societal structures and content, in relation to its own cultural and historical forms and ideals. In this way we can critically judge whether "actually existing democracy" is the result of a distortion or suppression of its own historical and cultural potential for "genuine democratic participation", without importing or imposing forms and ideals upon that culture or country.[31] The critical judgement of whether this potential is being realised or not can be made by examining the meaning of "democracy" in the context of its use and reception, in relation to whether it involves a continuity with its historical or cultural tradition, whether it respects and improves upon that tradition, or whether it is a comfortable fiction within a society that declares itself to be democratic, while being rife with social inequalities, corruption, electoral manipulation, and media control. Regardless of its history, we cannot consider a nation-state to be democratic if it has an authoritarian government that overtly exercises police-state tactics and propaganda campaigns; overrides the law either by decree or executive privilege; monitors, arrests, imprisons, and even tortures political opponents without any regard for due legal process; and, conducts militaristic acts of state-terrorism over foreign countries and cultures in order to protect or intensify the privileges, wealth, and power of its economic elite. If the idea of "democracy promotion" is to have greater meaning than being an empty slogan, then it needs to be critically and conceptually understood in terms of its relation to its own historical tradition and how it is used within the social ontology of practices and discourse about how citizens understand themselves as the *demos*, their potential and limitations for participation, and their political freedoms, rights, and duties as citizens.

The competitive elitist model of liberal democracy has resulted in a political system of compromises and exchanges between representatives of the economic elite, which reinforces the *status quo* and suppresses all efforts to develop a broader, rational understanding of how to realise the public good and well-being.[32] This is not just a matter of conspiracies between corrupt politicians and greedy capitalists, although clearly these do occur, but it an institutionalised consequence of the technical system of supplying funding to pay for private owned mass media time required for political campaigns in a mass society. It is the result of the intensification of the relation between capital accumulation, access to media, and the costs of re-election. Even if the owners of mass media remain impartial, due to the mass media costs of campaigning, candidates for public office are vulnerable for being selected on the basis that only those that are able to provide sufficient campaign funds are the only people who can afford to run for office. Candidacy is limited to billionaires or those that are able to attract donations from special-interest groups and, thereby, gain party support, access to mass media, and endorsements. Either way, such candidates will have a vested interest in preserving the *status quo* because only those candidates that either are members of the economic elite or serve the interests of the economic elite can hope to be elected to public office. As a consequence of this, the competitive elitist model of liberal democracy favours technocratic plutocracy within which the function of an election is limited to legitimating the *status quo* as being democratic, regardless of how unrepresentative the candidates actually are. The commonplace prejudice that "actually existing democracy" is "the lesser of evils" because some kind of centralised state apparatus is necessary to effectively solve societal problems flies in the face of empirical facts, such as the inability of powerful nation-states and "the invisible hand" to deal with either international problems (such as environmental pollution, poverty, warfare, child slavery and prostitution, the proliferation of nuclear weapons, energy crises, etc) or national problems (such as increasing levels of unemployment, crime, poverty, child abuse, traffic congestion, urban degeneration, health care costs, educational standards, etc). It seems more the case that the activities of the agents of nation-states actually undermine efforts to solve these problems or are the causes of such problems, while the nation-state is continually represented by its agents as the best means to protect "the national interest" and "national security" by increasingly resorting to the use of police and military powers, without regard for civil liberties, human rights, or international law, as a reaction to the consequences of international and national problems. In this respect, liberal democracy overly relies on a fundamental interest on the part of agents of the State to preserve the institutional arrangements upon which liberal democracy depends. This fails to attend to the way that these institutions are in fact transformed to serve the interests of the political elite that is responsible for the state apparatus "on behalf" of the electorate. As a result, the liberal competitive elitist model of representation is a substitute for democracy, rather than an approximation.

Participatory democracy, uncertainty and the good life

As I argued in *Modern Science and the Capriciousness of Nature*, we exist in a world within which we are subject and vulnerable to forces beyond our control, such as extremes of weather, earthquakes, volcanoes, disease, accidents, and, ultimately, death. Modern science and technology are deeply implicated in a societal struggle to pacify existence by using our knowledge of natural mechanisms to develop devices and techniques to alleviate humanity from the limitations of our material conditions. Science and technology are driven by the desire to achieve certainty and control in an often chaotic world that seems indifferent and is frequently hostile to human life; to overcome our sense of vulnerability to natural forces beyond our control by constructing an artificial world – a technological society – which promises to liberate and protect us from the destructive power of Nature. From their onset in the sixteenth and seventeenth centuries, modern science and technology have been implicated in a societal gamble upon the rationality and goodness of the technological society, as a substitute for the natural world, by confronting and appropriating Nature to transform it into a more intelligible and controllable world of our own making.[33] Modern science and technology comprise the Baconian response to Thucydides' ancient dream by promising to construct a paradise on Earth to liberate human beings from toil, scarcity, and suffering, and prevail over "the brute nature of things".[34] The Enlightenment project of constructing a rational society was premised upon the achievement of human freedom using our natural faculty of reason augmented by the knowledge, experiences, and powers discovered in mathematics, the practical arts, and natural science. This project was intensified during the nineteenth century, within which the Industrial Revolution led to the rapid development of science and technology, alongside the positivistic reduction of the conception of progress and rationality to instrumental reason and its application to the enhancement of efficiency, technique, innovation, and power. During the twentieth century, science and technology have been applied to nearly every human activity, including agriculture, education, information, labour, medicine, politics, sexual reproduction, and warfare, and increasingly requires large teams of researchers and technicians, high levels of funding, and is intimately connected with the so-called industrial-military complex.[35] The trajectories of scientific research and technological innovation have become almost indistinguishable and combined into technosciences directed to provide new devices and techniques to satisfy the demands of production and consumption.[36] Biotechnologies, including genetic screening, selection, and modification, as well as cloning, are the latest technoscientific manifestations of the societal gamble, promising to create a brave new world free from birth defects, hunger, and disease, within which human beings improve upon how life begets life, as well as farming and healthcare.[37] However, the argument for participatory democracy presented in this book depends upon what Bruce Bimber termed as an "unintended consequences account" of technological development, which highlights the fact that the consequences of implementing and developing technology in the world are not determined by human intentionality, but, rather, that technology has its own agency and its development is uncertain, unpredictable, and uncontrollable.[38] However, my argument deviates from technological determinism because, as I shall argue in the next chapter, a substantive theory of technology explains how technological development shapes the organisation of society, including human choices, actions, expectations, and intentions, the relations of production, including political and economic institutions, and patterns of consumption, without appealing to either natural or historical laws. In Chapter 3, I argue that human choices and judgements, in the absence of universal consensus of the best course of action, dialectically shape the trajectories of technological development and, in turn, are shaped by them. In many respects, I critically develop what Bimber termed as normative technological determinism by showing that a substantive theory of technology offers the basis for a critical departure from normative technological determinism by rejecting the dominance of technical norms, instrumental reason, and technical rationality over human thinking, deliberation, and decision-making, while showing how human choices are shaped by technological activities. In this respect, the argument of this book follows on from "the humanist tradition", as Carl Mitcham termed it.[39]

Unintended consequences accounts are actually premised on technological indeterminism or underdeterminism. Technological development is indeterminate in the sense that the interactions and interconnections that emerge from the processes involved in integrating new technologies into pre-existing complexes and ensembles of other technologies are themselves bounded by human choices and judgements, dialectically modified to accommodate to the resistances and demands that are created in response to the processes of integration, refinement, and further development. They are the product of heterogeneity and inconsistencies in non-linear interactions, within an incomplete technological framework, compounded by the latitude inherent to human decisions made in the face of the irresolvable ambiguity regarding the consequences of our actions. Technological development is underdetermined in the sense that the totality of possible uses, as means to ends, of any device, tool, or technique is an open set, and, therefore, an unintended consequences account simply accepts the veil of ignorance about the future, the epistemological and ontological limitations this implies about the relation between technical expertise and the knowledge of how to understand and realise public goods, and human fallibility. It is for this reason that I deny that technical expertise necessarily conveys any special insight into the consequences of technical innovation, and, therefore, given the absence of any universally accepted epistemological or moral standards by which we should choose or judge the best course of action, it is prudent to hedge our bets, so to speak, and include as diverse and pluralistic a stock of experiences, knowledge, values, ideals, and skills, as is humanly possible. The societal capacity to discover human goods is optimised by diversifying human experience and providing the means to communicate those experiences, and maximal democratic participation increases society's capacity for plurality, diversity, and creativity. My argument is that the health and sustainability of society depends on participatory democracy because this improves the societal capacity to incorporate diverse and pluralistic knowledge, skills, values, and ideals, within the societal structures for communication, deliberation, and decision-making, which increases the societal level of creativity, adaptability, and experimentalism, while allowing local communities to preserve local traditions if these are deemed satisfactory by local citizens. Democratic participation creates an evolutionary dynamic, through encouraging dissent and criticism, which tends towards experimentalism. At a local level, any tradition (including religions, sciences, laws, ideology, values, and philosophies) can be affirmed as providing the framework for community organisation, but at a societal level the legitimacy of each tradition is brought into question in the light of the plurality of available traditions. This simultaneously resists the tendency for one tradition to dominate, encouraging different traditions to be explored in different communities, and this recovers the potential for an emerging consciousness of the experimentalism inherent to all traditions, while highlighting the inherent pragmatism at play in choosing between possible practices. It is thereby essential that the trajectory of democratic participation emerges from the particular to the general, so to speak, through the exploration of various traditions and experiments in different contexts and localities. This is not so much a model of bottom-upwards democracy, but of local-outwards democracy, achieving consensus through horizontal enrolment, rather than democratic centralism, which in any heterogeneous society will constitute associations and alliances between confederated communities, workplaces, movements, and organisations to realise shared visions and address over-lapping concerns.

The case for participatory democracy should not be couched in terms of civic responsibility or moral duty, but, rather, as a process of practical value that would significantly enrich the lives of participants. It is based on the argument that "actually existing democracy" is the result of the distortion and suppression of democratic participation, which leads to only a small minority deciding the criteria and directions for societal development, and reduces the creative capacity of society to adequately respond to unforeseen events in an open-ended, changing, and complex world. It is not enough to show the shortcomings of "actually existing democracy". The practical value of democratic participation can only be realised if people have a high level of confidence and trust in democratic participation as a public good, which depends upon its success in solving practical problems for communities and workplaces. It must be tested through particular experiments. Once democratic participation becomes the means to solve many of the practical problems of everyday life, the stronger will be the public commitment to participatory democracy, and people can remain committed to it even in the expectation that sometimes it will result in mistakes, providing that the process remains open to their participation to correct those mistakes. Even when an individual disagrees with the majority, s/he will remain committed to the process of collective decision-making and continue to try to persuade the majority to change their view only if the process has practical value and the individual's participation can change the outcome of process. In this way, individuals will have trust in participatory democracy, even on the occasions when they do not agree with its collective decisions, because they expect to have genuine opportunities to change those decisions through future participation. If participatory democracy was impractical and prohibited dissent then it could only be sustained through coercion, which would further erode the trust and confidence in the process, and it would cease to be a participatory democracy in any meaningful sense of the term. The practical value of democratic participation is dependent upon the level of creativity and coordination that can be achieved through cooperative and critical deliberation between all the concerned participants. The democratic process must be the best means to discover and satisfy common goods and ideals, which were previously neglected or unknown. Hence, democratic participation should not be considered as a means to resolve conflicts or for achieving a compromise between individuals, but, rather, as a creative, social process of bringing together dissenting voices, diverse skills, ideas, and experiences, as well as a plurality of goals, ideals, and values, in order to discover new possibilities and the best course of action to achieve them. Participatory democracy not only opposes "democratic centralism" or "majority rule", but encourages dissent and criticism to continue even when a consensus is reached. It also rejects any notion that the agenda should be set before the deliberative process begins. It is not only futile to attempt to establish any single definition of democracy, given the plethora of such definitions, but it would also pre-empt the public deliberation and decision regarding how best to organise and conduct itself as a democratic process. At most, one can only point out the minimum conditions for such a process to begin and maximally include citizens in the decision regarding how the process should continue towards forming a *demos* capable of self-governing by providing all citizens with equal opportunities to participate in the deliberative and decision-making processes that directly affect their communities, workplaces, and the public dimension of their lives. However, the rules, procedures, and outcomes of such processes should remain a matter for those involved to resolve for themselves, as they deem appropriate and acceptable, given that they expect to experience the consequence of their decisions. Even though the democratic process cannot guarantee good decisions, it is necessary because there is a lack of universal consensus about what the best course of action should be, given an absence of universally agreed epistemological and moral standards, and, therefore, what qualifies as a good decision remains ambiguous, contested, and at stake as outcomes of the democratic processes involved in deliberating and deciding the best course of action. Participatory democracy is a process of social evolution given political form that would emerge in response to a shared understanding that the consequences of our actions are not determined by our intentions, alongside a shared commitment to achieving the good life and well-being for as many people as is humanly possible, to be achieved by including all the people for whom the outcome of any proposal is a matter of direct concern. It does not require that we all share the same understanding of reality or the same con-ception of our obligations to each other. While we may well accept that some solutions will be better than others, to any given problem, in the absence of any consensus regarding our assumptions, all proposed problems and solutions must be held to be worthy of consideration. If the aim is to discover the best course of action, we should not exclude any possibilities *a priori*, nor should we assume that every possibility will result in equally desirable consequences. Given that there is an absence of any universally agreed principles and methodologies from which we can derive the practical value of any proposal, it becomes a function of the process to evaluate proposals through criticism and persuasion between the people who are likely to experience the consequences of whatever course of action they decide to undertake. Hence,

it is essential that participatory democracy is egalitarian, but not rela-
tivistic; otherwise it would not have any practical value as a process of
discovery of the best course of action. In an open-ended, changing,
and complex world, the goodness of any course of action is underdetermined because all of its consequences have yet to occur. The
process of deciding the best course of action is necessarily one of ongoing
refinement, re-evaluation, and criticism. Furthermore, we also need
to adapt and modify the best course of action to accommodate unforeseen consequences, events, and circumstances. Nor should we presuppose that there is only one best course of action. Hence, democratic
participation should be analysed as an unfolding, developmental, educational, and experimental process.

John Dewey argued that the modern "quest for certainty" is actually
based upon "man's distrust of himself", which produced an imperative
to obtain the state of transcendence to complete and eternal knowledge.[40] However, once we accept our innocence regarding knowledge
of "the good", but we agree on the desirability of such knowledge, then
we can recognise the benefit of a pluralistic and diverse society, as an
experimental and open society, allowing different paths of societal
development to coexist, allowing different conceptions of "the good"
to be explored in the development of local communities and workplaces, including the implementation and further development of
scientific research and technological innovation. In such a society, the
nature of "goodness" is open to question, any proposal is equally worthy
of consideration, all standards can be challenged and contested. In this
way, a broad understanding of rationality can be achieved through the
inclusion of diverse and pluralistic criticisms and dissent. The creativity, practicality, and sustainability of society are maximised providing
that members from different communities can communicate freely with
each other and it is possible for people to learn from the experience of
others. The answers to questions regarding the goodness of any vision,
ideals, or proposed course of action may well remain contested, but
they have value as a focus for the ongoing critical examination and
deliberation of our potential in comparison to the actuality. It is on the
basis of such a comparison that we can make judgements regarding the
goodness of our goals and the rationality of the course of action proposed to achieve them. Once we accept that human nature is complex
and pluralistic, and that there is an absence of any universally accepted
description or explanation of human conduct, in scientific, religious,
or philosophical terms, we should not rely on any "meta-narrative" to
define the democratic process. It is for this reason that theorists should

avoid proposing overly determined or structured models of participatory democracy. While some account of "good conduct" and the nature of deliberation and decision-making will be assumed by every theory of participatory democracy, if it is to be an adequate theory then it should not overly rely on this account; instead, it should leave space for alternative accounts to coexist, even when such accounts are incompatible or incommensurable. The challenge is how to proceed with an enquiry into "good conduct" and "the democratic process" in the absence of universally agreed epistemological and moral standards. Hence, in many respects, participatory democracy depends on a Socratic development of intellectual virtues in the absence of certain knowledge. However, this involves challenging the elitist interpretation of Socratic enquiry, as made by Leo Strauss, for example.[41] He ignored Socrates' claim in *The Apology* – that we all have the potential for understanding within us, to be brought out by philosophical dialogue – in favour of the oratory of Socrates in *The Republic* – that the good society requires the guardianship of philosopher-kings, the intellectual elite who know the forms of truth, justice, and goodness. Of course, we may well agree with Strauss that to judge soundly one must have the correct epistemological and moral standards, but this agreement does not help us discover what those standards are. If the quest for truth, justice, and goodness involves ascending from shadows of the cave of the *hoi polloi* into the natural light of the Sun, we still need to know how to embark on such a quest. If we take seriously Strauss' claim that the task of political philosophy is to truly know the nature of political things, we still need to know how to found the epistemological or moral standards upon which to discover such knowledge. It is without doubt that it would be good for future legislators and statesmen to be educated by political philosophers who truly know the nature of political things, but how could we recognise such masters and test their claim to knowledge? Ironically, Strauss' rejection of relativism and historicism was based on a culturally relative and historically conditioned appeal to a "classical" interpretation of the Platonic philosophical tradition and a reliance on the soundness of his own naïve intuitionism, and, as a result, Strauss remained tied and gazing at the shadows on the cave wall. Strauss' political philosophy is premised upon little more than an intellectual elitism based on the assertion of an arbitrary set of cultural and historical norms as universal epistemological and moral standards, but it is unclear how he could justify these standards. However, if we return to the Socrates of *The Apology*, we may readily interpret the Socratic task differently. Thinking and reasoning are activities that are available to most (if not all) human beings, as being fundamental to their ontological realisation as human beings, rather than some kind of "expertise", providing that human beings learn to philosophically examine and test their opinions. Socrates distinguished between opinion and knowledge, given that the latter emerges from the examination of the former, but, as an improvement of opinion through testing its soundness and consistency. Socrates remained a citizen among citizens, making no claim to wisdom or knowledge, hence, refusing to accept payment, as any sophist would. Even when facing his (unjust) death, he maintained his claim to be a citizen among citizens, accepting the judgement of his fellow citizens by refusing to allow his friends to help him escape to another city. Hence, if we take Socrates in this light, as a citizen among citizens, we can interpret his oratory in *The Republic* as showing the importance of philosopher-kings to guide the education of the citizens in the good society, but also showing their absence from Athens and, thereby, highlighting the importance of democracy, as the next best form of governance, with all its limitations. In the absence of Plato's philosopher-kings, there are not any experts when making decisions about the good life and we are all on a par, as amateurs with a vested interest in discovering for ourselves how to live "the good life" and its relation with "good conduct".

The argument in this book opposes the elitist and positivistic tendencies within political science to limit itself to "empirical" descriptions or analyses of the current state of affairs, while asserting epistemological and technical standards in order to ascertain and analyse "the facts". Positivistic political scientists tend to reject theories of participatory democracy as being normative and value-laden. These political scientists argue that political science should take as its object the empirical facts of political institutions, relations, and events, and the meaning of "democracy" should be limited to "actually existing democracy" by mapping out how the word "democracy" is currently used, rather than given meaning in relation to the norms, ideals, and values of political philosophers. However, the argument presented in this book is that advocates of participatory democracy should reject the positivistic assumption that only a purely descriptive and analytical political science is desirable and rational. Such an insistence is both naïve and unhelpful. Political philosophers need to recover the normative assumptions and ideals that underwrite the interrelated representations of the human condition and potentiality that are inherent to all political theories and manifestos. They also need to show the constructivism inherent to positivistic political science. All political philosophies and sciences are normative and reflect cultural meanings, presuppositions, and pre-judices; political philosophy should reveal these normative and contin-gent aspects of positivistic political science and deconstruct its historical claim to objectivity, value-neutrality, and impartiality. Positivistic polit-ical science has a long history of development of the moral and episte-mological standards required to allow the construction and utilisations of historically conditioned and contingent representations of the human condition and nature to construct methods of analysis, measurement, and selection, under the guise of scientific rigor and realism. It is on the basis of these methods that positivistic political science is able to construct "the facts" and present itself as an empirical science (on a par with the natural sciences) capable of discovering underlying tendencies and law-like structures of political events and activities. It is only in virtue of its utilisation of shared cultural meanings, representations, and prejudices that its status as a science is secured within contem-porary culture and its results can be represented as universal truths. However, as Hans-Georg Gadamer argued, once we recognise that all representations (including those used to guide us in how to construct and apply representative techniques) imply interpretations that pre-suppose sets of beliefs, norms, standards, and conventions, the integral part these assumptions have in theoretical developments is itself an integral part of how we understand communication and rationality.[42] The methods we use to select the important aspects or characteris-tics of experience, or how we use historical interpretations to construct those representations or explanations of experience, actively transform how we go about selecting these aspects or characteristics. The method-ological ordering of experience, required for the identification of the facts, utilises representations and interpretations that are brought to the experiences to unify them as experiences of "something" that tran-scends those particular experiences and categorically and conceptually relates them to others. Positivistic political science is based on the projection of methods of representation and interpretation over exper-iences, which modifies and refines those experiences, in order to produce an intelligible and communicable fact. The claim that this projection reveals value-neutral data presupposes that the methods are also neutral. However, such a claim cannot be based in experience, given that one cannot experience "neutrality" independently of value judgements, and, therefore, positivistic science presupposes epistemo-logical standards that are not themselves empirically grounded. It depends on the presumption that epistemological standards can be deduced from the current set of representations and interpretations of human beings and the natural world produced from scientific activities. In this regard, positivistic science is an experimental science of imposing its own interpretations, interventions, and judgements, in accordance with its assertion of the epistemological standards of the natural sciences, which makes it as normative as any political philosophy and it is as guilty of "piecemeal social engineering" and being a "pseudo-science" as Karl Popper accused Marxism or Freudian psychoanalysis of being.[43] Furthermore, by trying to model political science on a positivistic conception of the natural sciences, a crucial aspect of political philosophy has been lost. There are aspects of political life that involve making interpretations, judgements, and choices in response to irresolvable ambiguity. These cannot be reduced to statements of preferences or estimations of necessity because they are only meaningful when we are confronted with ambiguity and uncertainty, while the outcomes of our decisions have vital importance for the quality of our lives. Inducing political decisions from the outcome of some method is inadequate because the quantification of preferences or the likelihood of consequences cannot provide us with certain and objective knowledge of "the good" or "the best course of action". It is an essential task for political philosophy to understand how we should deal with ambiguity and uncertainty, when making normative decisions regarding "the good" and "the best course of action", and, unless political philosophy addresses how we should make these choices, judgements, and interpretation then it has very little to contribute to the understanding of the purpose of political activity. Without any such contribution, we have little need of political science at all, except as a technical means to stabilise an arbitrary political system and derive policies from the statistical analysis of irrational preferences.

It is thereby essential that we try to be as explicit about our ideals and beliefs, as is humanly possible, as the interpretive basis for our theoretical developments and critiques. Normative theories of democratic ideals provide a critical position against which "actually existing democracy" can be evaluated. In order to do this, we need to explicate the visions of the good life, the ideal society, and human well-being that such normative theories entail or presuppose. We need to reveal and challenge the implicit judgements and visions about human values, goods, and goals presupposed in political decisions regarding the directions of scientific research and technological development. It is for this reason that I shall critically develop a critical theory of scientific research and technological development as being historically conditioned and contingent upon human attitudes, while also developing the call for democratic participation to place bounds on "the technical" in order to subordinate it to lifeworld experiences, meanings, and choices about how to explore and discover visions of human well-being, the ideal society, and the good life. We need to move beyond consideration of whether and how the public can become more involved in decisions regarding the directions, moral limits, and funding of scientific research and technological innovation. This book is not concerned with the philosophical problems involved in specific ethical and political questions about the public legislation and regulation of science and technology. Of course, these philosophical problems are important for the public understanding of the limits of and possibilities for making demands and placing constraints upon the directions of scientific research and technological innovation, especially concerning contentious technosciences, such as biotechnology or nuclear physics. Consideration of these problems is also important for questions regarding the social justice of access to scientific knowledge, the distribution and directions of funding scientific research, and the decisions regarding how science is to be organised within society for the public good, and many other equally important questions.[44] However, while I acknowledge the importance of these questions, I am not concerned with promoting an external imposition of democratic values upon science. Of course we need to critically examine the rationale and justifications for limiting and directing scientific activity in accordance with political and ethical considerations, but the argument in this book takes a step back from these important questions and considers the more general philosophical question of the meaning of democratic participation in the context of the substantive development and implementation of scientific research and technological innovation. This book is concerned with the question of the possibility of identifying and recovering *internal democratic values* within the technical and cognitive basis of scientific activity, which have been suppressed or distorted within contemporary society, affecting how science is done and understood, and how the recovery of these values will improve the quality and stability of scientific research and technological development. The argument given in this book does not assume that democratic participation is a self-evident *intrinsic* societal good and moral right. If we are concerned with the undemocratic character of modern society, we need to ask harder and deeper questions than asking how technology and science could best be developed in order to satisfy this societal good and moral right. While I sympathise with the assumption that democratic participation is an intrinsic good, my argument is that it is also crucial to make a strong case that *direct democratic participation is an instrumental societal good because it increases the chance of the rational and sustainable development of science and technology, which increases the chance of the rational and sustainable development of a scientific and technological society.* In this book, I shall present such a case for democratic participation.

2
Substantive Theories of Technology

The Ancient Greek poet Sophocles warned us that we will become increasingly dependent on the artificial world that we construct to liberate us from our physical limitations and protect us from "the evils" of the natural world.[1] Sophocles lamented the tragedy of the human condition, existing in a cosmic order that is indifferent to human suffering, for which *techne* (artifice) is the means to escape suffering by conquering Nature through agriculture, medicine, mechanics, and architecture. This allows human beings to live comfortably in cities, developing politics and the other arts, but, ultimately, humanity will become dominated by its dependence upon artifice, which perpetually creates and destroys, as it drives human beings to innovate an inhumane world that is beyond human control. It is this fundamental concern with the substantive impact of technology on human freedom and the human condition that has been a recurring theme in the damning critiques of "mass society" offered by many twentieth century philosophers and social theorists, including Martin Heidegger, José Ortega Y Gasset, Hans Jonas, Hannah Arendt, Lewis Mumford, Eric Fromm, Hans-Georg Gadamer, Walter Benjamin, and Jacque Ellul. Carl Mitcham termed this as "the humanist tradition" of the philosophy of technology.[2] Max Weber argued that, in an advanced industrial society, political decisions are constrained and directed by technocratic systems of calculation, assessment, and control.[3] Although Weber considered sciences to be value-neutral, in the sense that they cannot tell us what to do or how we should live, he argued that the development of modern science has increasingly limited our freedom to pursue alternative courses of action.[4] Politics is no longer the art of the possible. Political judgements about the best course of action presuppose technical judgements about what is possible and how best to achieve it.

The identification of "the possible" has become the jurisdiction of technology and, therefore, as Ellul argued, "the political" has become subordinated to "the technical".[5] Technology has radically transformed human agency and experience. It has transformed society and the natural world in irreversible and unpredictable ways to construct an artificial world – a technological society – wherein human beings hope to be liberated from toil and suffering. Science promises to provide us with the means by which we can understand the means to construct this technological society, by providing a better understanding of our material conditions and discover new "forces of production", in a way that is culturally represented as the organisation and application of natural mechanisms, but science is itself a technological activity that represents Nature in technological terms.[6] Science and technology transform, condition, and constrain our expectations about the modal necessities, possibilities, and limits of our agency, which, of course, shapes how we understand the material conditions of our existence. To the extent that we define ourselves in terms of what we do, our technological activities, at least in part, define who we are. They structure how we participate in society, shaping the horizon of possibilities towards which we aspire, providing that we conform to the discipline they demand. Substantive theories of technology show how technological development structures, directs, and constrains human agency, intentionality, and experience. In this sense, technological innovation causes social changes and adaptations, as well as creating new possibilities for further development, shaping human values and goals, but technological development is itself directed in accordance with human choices, attitudes, and expectations. Substantive theories of technology reject the assumption that technology is a neutral means, developed in accordance with either natural or historical laws, and also reject the commonplace notion that human intentionality is independent from the means at our disposal. Technical judgements are conditioned and contingent modes of ordering and organising alethic modalities that endure from one generation to the next. Human thinking and planning have become scientifically and technologically mediated and bounded at every level as the result of historical choices. Human goals and practices are selected and structured in accordance with the necessities, possibilities, and limits of the technocratic system through a process of "reverse adaptation", as Langdon Winner termed it.[7] Once a device or technique is implemented as "the most efficient means" available, technological imperative to use "the most efficient means" imposes a demand upon human beings to use them, while the commercial and political use of propaganda disseminates the idea "the most efficient means" are well understood and necessary means to achieve well understood and necessary ends, in accordance with the objective, rational, and logical trajectories of scientific research and development. It is in this sense that modern society is perpetually developing new ill considered means to ill considered ends and shows how technical rationality can dominate human intentionality and suppress critical reasoning to such an extent that technological innovation becomes an irrational end-in-itself. As Karl Manheim argued, if taken to its irrational extreme, the assertion of technical rationality can suppress, distort, or damage the human capacity to use our intelligence and insight to critically evaluate the desirability of postulated ends and the proposed means to achieve them.[8] Critical Theorists, including Theodor Adorno, Max Horkheimer, and Herbert Marcuse, developing and refining the application of Marxism and Freudian psychoanalysis to "mass society", have expressed deep concerns with the modern tendency towards the construction of a totalitarian society. They were concerned with the substantive role that positivistic science and technocracy have in constructing a technological society, wherein human thinking is reduced to instrumental reason and calculation. Critical of both industrial capitalism and state-socialism, these theorists were deeply concerned with the modern tendency towards irrationalism (barbarity) and also conformity to technical systems of production and patterns of consumption. Inspired by "the humanist tradition" and Critical Theory, contemporary social theorists, such as Jürgen Habermas, Langdon Winner, Andrew Feenberg, and Richard Sclove, have developed substantive theories of technology, as well as others, including Thomas Hughes, Andrew Light, Donna Haraway, and Albert Borgmann, to help us critically reflect on the substantive impact of technology on human freedom and experience, as existential and sociological phenomena, taking seriously the structural interrelations between politics and technology, and critically examining how technological innovation effects the conditions for the rational development of society.

Technological determinism

Substantive theories of technology attempt to explain how historically developed and structured technologies constrain and direct human choices and practices, while also rejecting technological determinism, which holds that society is the product of technological development and the trajectory of this development is governed by historical or natural law. Anarchist and Marxist theorists have often assumed technological determinism in their accounts of the current state of affairs and how a libertarian, egalitarian, and rational society can be achieved. Pierre-Joseph Proudhon, Michael Bakunin, and Karl Marx represented science and technology as the objective and rational means for human liberation from superstition, toil, and suffering by dominating the natural world, discovering new forces of production, and, driven by the technological imperative to improve productivity, empowering the future construction of a technological society as a socialist post-scarcity society.[9] Under the technological imperative, inherently enhancing the potential for human emancipation, the aim of technological innovation is to substitute automated machine performances for all human labour. The historical struggle to construct the technological society as a post-scarcity society promises to culminate in the creation of an egalitarian and integrated infrastructure of automated relations of production and distribution subordinate to human will. The fundamental relationship between human beings and the material world will be transformed from labour into design. By liberating human beings from scarcity, technology has the potential to liberate the creativity of human beings by allowing time for education, cultural activities, and leisure, but, technology has been historically conditioned by scarcity and exploitation. When technologies have been built to oppress one class of human beings for the benefit of another, the actuality and potential for oppression is enhanced. This effects the future development of technologies and leads to an increasingly sophisticated and powerful trajectory of the research and development of means of oppression. Hence, both anarchists and Marxists have called for a revolutionary seizure of the means of production in order to recover the liberatory potential of technology. Anarchism and Marxism simply differed on their conception of the role of the State in this revolution, which for the anarchists was an inherently oppressive organ of power; for the Marxists, the State ("the dictatorship of the proletariat") was historically necessary for the transition from capitalism to communism. The idea of "historical necessity" and the demands of the technological imperative, as revealed by scientific knowledge, were represented as one and the same under the Marxist conception of the socialism and the construction of the post-scarcity society, wherein the State would "wither away" and its functions absorbed by civic society. Adherence to the demands of technical rationality became represented as a moral duty towards the future of humanity in general; and resistance or opposition was to become represented within Soviet ideology as the product of backwardness or bourgeois idealism. Intensified and distorted under the Bolshevik nationalisation of industry and the Stalinist agrarian collectivisations, the Revolution became identified with the technological imperative to maximise productivity and any resistance or opposition was represented as counter-revolutionary.

Indeed, Marx stated that "technology discloses man's mode of dealing with Nature, the process of production by which he sustains his life, and thereby also lays bare the mode of formation of his social relations, and the mental conceptions that flow from them". [10] Marx considered technological activity to be the condition for objective scientific knowledge and assumed that it has an emancipatory essence to liberate human beings from labour, which has been perverted by capitalism for the sake of increasing profits by reducing wages and intensifying labour, while increasing productivity and lowering costs. [11] He argued that social and economic relations of production were the products of history and not the laws of Nature, but he also accepted the objectivity of the laws of Nature and their effect on our organic and material conditions for existence. Human beings are distinguished from other animals because we produce our means of existence, a step which is conditioned by our organic being, but not determined by it, and, by producing our means of existence, we are directly producing the conditions of our existence and our history. Technology does not just provide the means by which we manipulate and control our material conditions, but it also provides the means by which we consciously realise our essence as human beings. As he put it,

> The mode of production of material life conditions the social, political and intellectual life process in general. It is not the consciousness of men that determines their being, but, on the contrary, their social being determines their consciousness. [12]

He considered natural science to be a theoretical relationship between human beings and Nature that was internally bound together with practical activity, industry, and with the development of "the forces of production". However, economic class has lead to a division of labour which abstracts science from this internal bond and suppresses a conscious recognition of it, while developing the scientific understanding of objectivity in accordance with the values and ideology of capitalism. Once we recover a consciousness of this internal bond, we can understand how the scientific understanding of Nature has been developed through practical activity, industry, and the development of "the forces of production", alongside a scientific understanding of how human beings have evolved from our organic conditions by transforming the material conditions of our existence. The aims and means of scientific research have been transformed through the material practices of previous generations, creating new instruments of observations, new experiences and refinements, which allow new theoretical representations and conjectures to be possible. The theories and practices of science have been developed within the bounds of "the relations of production" at any given stage of the development of "the forces of production" and, thereby, cannot be abstracted from these bounds without distorting our understanding of them and hindering the development of scientific objectivity. It is through the conscious engagement with reality through material practices and sensuous activity that provides science with its objectivity, while the understanding of that engagement is itself developed and differentiation within a dialectical, historical process of transforming "the relations of production" in order to resolve the contractions generated by integrating new "forces of production" into these relations and the subsequent necessity for their further development and differentiation. Science and technology are thereby conditioned by the historical stage of development and differentiation of "the relations of production", within which they emerge, but they also have an intrinsic potential for the progressive transformation of society. Once society has achieved a sufficiently advanced level of technological development, science will transform itself from being a means for the exploitation of the proletariat into a condition for the emancipation of humanity as a whole. Exploitative "relations of production" will be inadequate and inefficient for the integration of the new "forces of production" into society. At which point, science will rationalise society by providing the means of human mastery over "the forces of production", alongside the conscious awareness of human essence as free and creative, which will demand that the inadequacies and inefficiencies of "the relations of production" are resolved in a way that satisfies human mastery, freedom, and creativity. For Marx, technology provided the site for the dialectical relationship between the historical development and natural evolution of human beings as social beings.[13]

The conditions for the emergence of a disciplined and class-conscious proletariat, as the agent for revolutionary change, are identical with the conditions for the unfettered construction of the technological society. It is the demand to overcome the limits placed upon the technological imperative by capitalism, in accordance with its demand for an adequate return on investments, which will lead to the overthrow of capitalism by a disciplined and class-conscious proletariat. According to Marx, the revolutionary seizure of the means of production could and should only occur once science and technology, within industrial capitalism, have reached a barrier to their further development, due to the limits of capitalist economics, and a technically advanced, disciplined, and class-conscious proletariat constitutes the means for the production of societal wealth. The revolution could only succeed once the proletariat have become capable of liberating the technological imperative and trajectories of scientific research from the limits of capitalist economics. It is in the sense that capitalism must have fully matured and reached its own developmental limit before the seizure of the means of production by the proletariat becomes historically necessary to overcome the limitations that capitalism imposes on the construction of the technological society. Once the proletariat achieves an advanced state of discipline and class-consciousness, identifying itself as the organised agent of productivity, then the demise of capitalism is inevitable once its technical limit has been reached and the proletariat consciously takes upon itself its historical task to construct and complete the technological society. In this sense, the emerging class-consciousness of a disciplined proletariat should not be understood as a fraternal sense of moral outrage at the privileges of the bourgeoisie, but, rather it is the conscious realisation of the necessity of the liberation of the technological imperative from the limitations of "the relations of production" inherent to capitalist economics for the societal construction of the technological society. On this account, "the dictatorship of the proletariat" would be the total societal submission to the technological imperative as being the only legitimate basis for the societal development of a rational, post-scarcity society.[14] Marx's vision of a communist society was indeed that of a post-scarcity technological society, wherein societal development would be directly controlled by scientists, technicians, and producers, comprising the vast majority of the population of such a society. Human beings would be liberated by technology from material needs, toil, and struggle, which would offer the freedom for the pursuit of education, leisure, and creative activities. After abolishing the capitalist perversion of its essence for the benefit of the minority, the technological imperative would construct the technological society for the benefit of the vast majority. Hence, for Marx, revolution would inevitably result from the contradictions between the ongoing technological innovation of "the forces of production" and the inefficiencies of the "relations of production", providing that society has achieved a state of advanced industrial capitalism, within which the vast majority of the population would comprise a disciplined and class-conscious proletariat. The revolutionary seizure of the means of production and the transition to "the dictatorship of the proletariat" would be the rational, technical means to construct a post-scarcity society, the end of struggle and inequality, the absorption of the administrative functions of the State into civic society, and the achievement of communism. Once the proletariat controlled the means of production – the machines at their disposal – the emancipatory essence of technology would liberate them to control the conditions of their existence. Upon completion, the technological society and the proletariat would be unified into a self-creating, self-conscious totality, wherein struggle and scarcity will be overcome. At which point, the proletariat would dissolve as a class into the conscious totality of the technological society and communism would have been achieved. It is in this sense that Marx's vision of communism presupposed the societal gamble in the goodness of the technological society and was necessarily totalitarian in its adherence to the dictates of the technological imperative.

It is important to distinguish between a totalitarian society and a society under dictatorship. The latter is a form of government that imposes its authority through the threat or use of violence, whereas a totalitarian society is one wherein alternatives or acts of resistance are impossible. As Hannah Arendt argued, totalitarianism involves domination of every sphere of human existence by means of the elimination of all human spontaneity and freedom.[15] While considerable violence would be necessary to create a totalitarian society, once it has been created, violence or the threat of violence would no longer be necessary. The totalitarian society would systematically control individual and mass participation, in accordance with an ideological vision of all the possible stages and destinations of historical development or human destiny, as it aims to transcend and direct all human potentialities by eliminating and suppressing all alternatives. While it is arguably the case that Marx's authoritarian tendencies were apparent during the 1872 split in the International Workingmen's Association ("the First International"), the problem with Marx's theory is not his authoritarianism, which may well have more inherent to Marx's character than his theory, but that the application of his theory would necessarily result in a totalitarian society, due to the presumption of the rationality of the technological imperative as the only progressive and emancipatory force for social development.[16] It is his assumption that technology is the only historical agent for change which presupposes technological determinism and a totalitarian vision of society.[17] Marx's vision was not a police state controlled by a central committee, wherein dissenters were arrested and sent to labour camps, but his vision was of a technological society wherein the technological imperative was inherently a moral imperative and the only rational basis for the egalitarian societal development of a post-scarcity society and the abolition of exploitation. Hence, Marx's theory does not provide any detailed discussion of how dissent and critical discussion should relate to the deliberation and decision-making processes of the majority, nor does it provide any theoretical basis for defending minorities (including ethnic, religious, and gender difference) and the rights of individual citizens. It was not simply the case that Marx considered these to be bourgeois ideas, but it was more the case that they were considered irrelevant, given that in "the dictatorship of the proletariat" only the question of how best to satisfy the demands of technical rationality would form the only rational basis for deliberation and decision-making. Such a society would be totalitarian because dissent and criticism of the technological imperative would be irrational and have nothing to contribute to progressive societal development, being represented as the product of backwardness and ignorance of the objective conditions for societal progress. Within any such society, it would become quite impossible to place any concerns into the public realm that did not deal with technological development or the relations of production and consumption, such as ecological or moral concerns that did not impact on the material conditions of human existence. Only science and technology would provide the conditions for the rational development of socialism into a communist society and, in this respect, Marx's vision of post-capitalist socialism was very different from "actually existing socialism".[18] The Soviet Union was the product of a series of distortions and suppressions of Marx's theory, which created an authoritarian and centralised bureaucratic administration of state-controlled capitalism that had very little resemblance to Marx's vision of communism and also distorted and suppressed the technical and scientific conditions upon which Marx's vision of "the dictatorship of the proletariat" and the transition from capitalism to communism depended.[19] The authoritarian tendency was apparent in Lenin and the Bolsheviks from the onset, which despite all of Lenin's rhetoric in *State and Revolution*, opposed any form of workers' control over the means of production.[20] If one compares Lenin's *State and Revolution*, with his previous writings in the *April Theses*, and his subsequent writings in *The Immediate Tasks of the Soviet Government*, published after the Bolshevik seizure of power, one can see that, apart from in *State and Revolution*, Lenin had little tolerance for worker's control, largely considering it to be anarchism. It is arguably the case that Lenin's "libertarian" and "democratic" views expressed in *State and Revolution* were designed to win over the workers' soviets to the Bolsheviks prior to the seizure of power. The fundamental ideology of the Bolsheviks after seizing power opposed workers' control and did not deviate from that expressed in Lenin's *What Is To Be Done?*, published in 1901, wherein he advocated nationalised and state-controlled industrial monopolies for the benefit of the people. From the moment they seized power in October 1917, Lenin and the Bolsheviks, as a minority with the support of the Red Army, imposed their vision of socialism upon the majority, in a way that required a series of distortions and deviations from Marx's theory.[21] Due to the absence of a disciplined proletariat and advanced means of production, the application of Marx's theory of socialism was contingent upon industrialisation, as was also the capacity of the Soviet Union to defend itself from military invasions by capitalist countries, which entailed the centralised planning of industry and agriculture in order to force the pace of industrialisation.[22] The growth of centralised bureaucratic administration and an institutionalised intolerance of dissent were quite inevitable under such circumstances and, therefore, Stalinism was not quite the aberration of Leninism that some Marxists would prefer to claim.

As a critical response to Marxism, Robert Heilbroner proposed a "soft" version of technological determinism that holds there to be an "inner logic" to technological innovation, but technological development is also directed by socioeconomic factors, which are strongly influenced by technology, without being completely determined by it, given that cultural and political considerations are also imposed on socioeconomic factors.[23] Heilbroner's sense of technological determinism was that specific stages of technological development have their own stratified technological conditions, i.e. the development of the hydroelectric plant required the development of the steam mill, which required the development of the hand mill, which required the development of basic gears, which required the wheel, etc. Each strata of technological development contains the previous strata as prerequisites, where Heilbroner attributed this "inner logic" to be conditioned by natural laws and mechanisms. The historical discovery of new "forces of production" is conditioned by the discovery of the natural laws and mechanism, but whether they are discovered or implemented is conditional on cultural and political influences, which cannot be reduced to either economic or technological factors. Heilbroner was aware that different cultures and societies have different goals and values, which lead them to develop different technologies in different ways, due to the considerable cultural and social pressures on the directions of technological development, but he argued that both the possibility of initiating a course of development and its continued evolution follow a determinate pattern rather than a random or contingent course. Heilbroner's position presupposed that technological development is based on the sequential and objective discovery and application of natural laws and mechanisms of Nature and, therefore, predictable and continuous, once initiated and sustained within any culture. Comparing Heron's and Watt's steam engines, Heilbroner argued that technological innovation is not simply a matter of producing a prototype, but also requires the successful integration that prototype within wider society as an effective and productive machine. This involves the establishment and development of a cultural infrastructure of supporting technologies (e.g. iron foundries and machine tools), as well as the discovery and dissemination of the technical knowledge of how to produce and reproduce the machine. It is this need for *technological congruence* that constrains the sequencing of technological development with the interrelation and connection between previously divergent and unrelated technologies that transforms a prototype, alongside the construction of its supporting infrastructure, into a new technology – this requires a transformation of social organisation and cooperation between different industries, technical specialists, and technologies within a stratum of organised and specialised knowledge, labour, and resources. New technologies only become "forces of production" when they are successfully integrated within relations of production, which, of course, changes previously stabilised relations of production. Once integrated into a new infrastructural stratum, creating new relations of production, new forces of production transform the social, economic, and political structures of the society from which they emerge, thereby changing social relations and societal organisation. The "inner logic" of technological development imposes constraints and directions upon the development of society by transforming the composition and division of labour and the hierarchical organisation of work, creating new societal needs and the means to satisfy them, as well as creating new forces of production for the further development of technology. Technology both reflects and moulds social relations; technological activity is a social activity that both responds to and imposes cultural values, directions, and influences. Any technological development must be congruent with the technological infrastructure of society and also compatible with cultural attitudes and influences. Hence, even though "the technical" is a subcategory of "the social", it cannot be reduced to it.

Bruce Bimber termed Heilbroner's theory of technology as a nomological variety of technological determinism because it presupposes that the conditions of technological development are governed by natural laws.[24] Substantive theory does not require any such presupposition, but, rather, critically situates the discovery of natural laws within technological development, as abstractions of that development. In *On the Metaphysics of Experimental Physics*, I argued that science is a process of transforming Nature into something that can be manipulated through technological activity, which is culturally represented as utilising natural processes without transforming them, while representing natural phenomena in technological terms and placing them at the disposal of technical apparatus and techniques of measurement. The conceptual contradiction between the artificial and the natural has been avoided by allowing the artificial to be metaphysically represented as being a natural consequence of natural human perceptive, cognitive, and reasoning abilities and capacities, within a material world, allowing the rational implementation of natural mechanisms in material practices to be simultaneously an act of discovery and practical application. In the case of experimental physics, scientific knowledge is simultaneously *techneic* and *epistemic* in the sense that it is characteristic of an objective asymptote of causal accounts, representing technical procedures in terms of natural mechanisms and laws, to explain the efficacy of techniques in the design, construction, operation, and interpretation of the performances of the apparatus in response to human interventions. Measurements cannot be simply attributed to the properties of the object measured, but must also be linked to the techniques used to make those measurements. It is for this reason that, given that scientific objects are defined in terms of "their" measurable properties, such objects are techno-phenomena and act as a substitute for the natural phenomena under experimental investigation. Scientific knowledge is an inherently metaphysical understanding of Nature, which represents natural entities and processes within a scientific world-picture as being the phenomenal manifestation of natural laws that govern the realisation and exercise of natural mechanisms that cause phenomenal changes through interaction, while it tests theoretical entities and predictions within the context of technological activity. Within this scientific world-picture, human being is represented as a natural being and an efficient cause that is capable of grasping the objective reality of its own material conditions by intervening into and manipulating those conditions by discovering and implementing "natural mechanisms" in material practices. The cognitive products of this effort are taken to be disclosures of the same underlying reality that causes the phenomenal natural world, human nature, and the "inner logic" of technological activity. The historical development of technology can be represented as the disclosure of deeper ontological strata, representing innovation as a mode of discovery, which can be unified as a progressive trajectory through modifications and refinements of the scientific world-picture, while the progressiveness of science can be verified in terms of its practical successes and instrumental value. It is only to the extent that an event can be situated, reproduced, and manipulated within a technological framework can it be said to be available to scientific investigation and, hence, taken to be part of objective reality. Only that which can be situated within the technological framework of experimentation and represented in terms of functions and consequences can be taken to be real. Experimental science aims to disclose our productive possibilities and explain these in relation to a scientific world-picture, rather than merely compare theory with experience (as the traditional philosophers of science would have us believe). It abstracts these productive possibilities into sets of alethic modalities, represented as corresponding to the objective reality via mechanical models, which are tested in terms of their instrumentality for future research, as well as for the practical problems of the wider world. The theoretical understanding of the process through which one set of technological objects are transformed into another set of technological objects is taken to be the theoretical representation of the objective natural mechanisms that permitted those transformations to take place. The ontology of the natural world has been substituted with a metaphysical interpretation of a collection of the performances of different kinds of machines and the epistemology of science has been reduced to a methodology of innovating, relating, and justifying procedures and techniques, by demonstrating their instrumentality in bringing forth novel technological power, and/or refining and developing the technological framework from which scientific research emerges. The scientific understanding of human limits, power, and capacity for control is identified, refined, and developed within this framework in order to present an understanding of ourselves and the world that offers both as being objectively within our cognitive and manipulative grasp. Only those aspects of the world that can be situated within the technological framework are understood as objective properties of the world, while the human abilities to confirm or refute theories, experiment upon and explore reality, develop and interpret new experiences, etc., are all situated, related, organised, and transformed within the technological framework. The refutation of scientific hypotheses and the collection of empirical data are only scientifically possible by remaining within the technological framework and, therefore, if this is the case, nomological technological determinism is based upon a positivistic presupposition of the neutrality of this framework that cannot be sustained on close inspection of how experimental science discovers natural mechanisms through technological activity.

The humanist tradition and the technological society

An early humanist critique of the substantive impact of science and technology on the human condition can be found in Jean-Jacque Rousseau's essay *A Discourse on the Moral Effects of the Arts and Sciences*.[25] By adopting a Socratic defence of virtue and goodness, opposing intellectual conceit and vanity, and declaring "the light of reason" as the means for the understanding of the universe and human nature, Rousseau considered the Scientific Revolution to be a restoration of the knowledge of the Ancient Greeks, after the barbaric Dark Ages, through "a complete revolution to bring men back to common sense". He considered cultural splendour and magnificence to be "the effects of a taste acquired by liberal studies and improved by a conversation with the world". However, when government and law provides for the security and well-being of human beings, the arts and sciences "fling garlands of flowers over the chains which weigh them down" and quell the inborn desire for liberty, causing them to become civilised and "love their own slavery". By satisfying the intellect and the body, the arts and the sciences empower governments to put themselves forward as the means to satisfy human needs and cultivate the talents upon which civilisation is based. By increasing "artificial wants", the arts and sciences further bind human beings to the social order, modify human behaviour, and provide artificial modes of expression for the passions. This leads to the refinement and abstraction of "our rude but natural" independence into "a servile and deceptive conformity". Human nature becomes expressed and restrained under the ornamentation of manners, such as politeness, affability of conversation, and social obligations to partake in the imperative to social benevolence. According to Rousseau, the development of the arts and sciences corrupts human nature because they pacify it, lead to vanity, and place powerful instruments in the hands of those ignorant of truth, justice, beauty, and goodness. Without such philosophical knowledge, we cannot know how to make the right use of scientific knowledge, nor even if it has any practical value at all.

By allowing the production of affluence and abundance, in the absence of philosophical knowledge, human beings are turned into an instrument of commerce and a resource for the State. Without a philosophical education to learn virtue and "the greatness of the soul", affluence and abundance lead to the increasing withdrawal of human beings into the private realm at the expense of the development and practice of public and civic virtue. By neglecting the development of a philosophical enquiry into morality, religion, and human well-being, education has been reduced to the acquisition of facts and skills, focussing on technical expertise at the expense of the development of a philosophical understanding of citizenship and happiness achieved through reason and "the voice of conscience".

The humanist tradition is the ongoing development of deep seated concerns with the oppressive and distorting character of modern society and its reductive and erosive effect on human freedom in both thought and action. Science and technology have undoubtedly brought great benefits, by providing techniques and devices to improve the material conditions of human existence, but it is the irrational conformity to the demands of technical systems that threaten to enslave humanity and erode the meaning of scientific knowledge. It was the positivistic subdivision of science into specialised technical disciplines that generated Edmund Husserl's despair regarding the state of scientific knowledge.[26] Husserl was deeply concerned with the extent that meaningful knowledge was becoming increasingly unavailable to European culture due to technical specialisation and increased levels of mathematical abstraction, which removed it from meaningful content in the lived world of human experience. For Husserl, theoretical knowledge should be intimately connected with the knowledge of how one should conduct one's life and he was concerned with the fact that the empirical sciences have nothing to say on this question. In his view, what leads to a scientific culture are not facts and technical skills, but the formation of an enlightened and meaningful understanding of life, but the positivistic presupposition of the importance of specialisation and technique had undermined this understanding of scientific culture and alienated human beings from scientific knowledge. This crisis was a result of the degeneration of one true science, a single meaningful *Wissenschaft*, into the advanced positivistic disciplines, such as physics, chemistry, and biology, which, in his view, lacked scientific rigor, as well as meaning, for the question of how to live well because they were unable to be unified into one true science through philosophical reflection. The cultural affirmation of the positivistic sciences was dependent upon historically conditioned judgements that had provided increased power but had lead humanity to a crisis. Martin Heidegger developed Husserl's critique further in *Being and Time* and showed that scientific knowledge was founded upon a special and specialised mode of being-in-the-world.[27] For Heidegger, modern science neglected to attend to the derivative character of theoretical knowledge from practical considerations and the central role of embodied human subjects in the constitution of meaningful knowledge from the lived-world of experience and practices.[28] Heidegger criticised the positivistic assumption that technique was prior to meaning and that the modern sciences had a privileged access to reality. The positivistic interpretation of experimental science as being a neutral and privileged mode of testing theories by using measurement and calculation fails to recognise the role that interpretation has in bringing data, results, and events into experience. It also fails to address the extent that phenomena are scientifically understood in terms of a framework of meanings that are assigned and refined through an ongoing process of interpretations derived from historically situated technological activity.

In *The Question Concerning Technology*, Heidegger was critical of the commonplace view that technology is merely an instrument to achieve human ends.[29] He considered this to be is the worse possible view of technology because it immediately places us in an unthinking relation with it. He was concerned with the question of how we can attain a free relation with modern technology by questioning its meaning and relation to truth. He was particularly concerned with the extent the technical substratum of our existence has become so immense that we are unable to cope with it and unthinkingly treat it as an end-in-itself. He argued that human beings have conformed to the technological imperative to gather and order "all their plans and activities in a way that corresponds to technology".[30] Nature and human relations are no longer valued unless they provide the means to achieve some carelessly considered end, which tends to be valued only in so far as it provides the means for something else. Heidegger considered modern technology to be distinct from ancient handicrafts because they constitute different modes of disclosure. Ancient handicrafts participated in *poiesis*, a mode of production that involves "bringing forth" beings into the world, as ends-in-themselves, which is related via *techne* (know-how) to *aletheia* (truth) by revealing the real through artifice in terms of completion and perfection.[31] Modern technology, on the other hand, is a mode of disclosure that sets beings in place – including human beings – and orders them to extract the release of energy for immediate use or storage. Technology sets upon land as a mining district and the soil as a mineral deposit to extract uranium, iron ore, or coal for some other use. Air is set upon to extract nitrogen for fertilisers for modern agriculture and rivers are set upon to supply water for irrigation or electricity generation. Nature is set upon and disclosed as a collection of resources and energy for future use. Technology is always

...directed from the beginning toward furthering something else, i.e. toward driving on to the maximum yield at the minimum expense. The coal that has been hauled out in some mining district has not been supplied in order that it may simply be present somewhere or other. It is stockpiled; that is on call, ready to deliver the sun's warmth that is stored in it. The sun's warmth is challenged forth as heat, which in turn is ordered to deliver steam whose pressure turns the wheels that keep a factory running.[32]

It is this availability for future use that Heidegger termed as "standing-reserve" to highlight its objectlessness as something for something else, without necessarily any consideration of what that something else would be. Modern technology discloses the world as pure instrumentality. However, technology can only disclose the world in this way because human beings respond to the challenge to exploit Nature and each other. Once human beings respond to this challenge, they must conform to the demands of the technological imperative and the discipline of the technological activity in question, whereby human beings are set upon by technology, gathered together and ordered into this mode of disclosure. He termed this as "destining", but insisted that this should not be taken to mean "a fate that compels... where 'fate' means the inevitability of an unalterable course".[33] Heidegger rejected technological determinism, but highlighted that, once human beings have accepted the challenge of technology to disclose Nature as standing-reserve, we do not choose or control what is disclosed as a result of setting-upon and ordering beings in this way. Destining is the human attitude of conforming to the technological imperative and the discipline of technological activity. Once human beings have submitted to destining, modern technology is a mode of enframing (*Ge-stell*) human and natural beings in relation to their instrumentality and "threatens to sweep man away into ordering as the supposed single way of revealing, and so thrust man into the danger of the surrender of his free essence".[34] Heidegger argued that this mode of enframing was first displayed in the rise of physics as an exact science. Experimental physics "is dependent on technical apparatus and progress in the building of apparatus" and it is a "way of representing [that] pursues and entraps nature as a calculable coherence of forces" and even as pure theory "sets nature up to exhibit itself as a coherence of forces calculable in advance, it therefore orders its experiments precisely for the purpose of asking whether and how Nature reports itself when set up in this way".[35] Physics prepared the way for the disclosure of Nature as a standing-reserve of energy available for future use by representing Nature as an ordered system that can be revealed through measurement, calculation, and mathematical description.[36] The materialism of modern science represents the world in terms of mechanistic interactions between inanimate matter and fields, thereby, reducing the reality of every being to its relation to labour and technological activity. Scientific truth, understood in terms of "correctness" (*veritas*), is perpetually deferred within the ongoing activities of experimental research and mathematical theorising, while Nature is represented as a product of mechanistic causes governed by laws that can be known in terms of the first principles of making changes in the material world (*techne*). Modern science has become the new metaphysics under the sway of the technological imperative that has found its positivistic culmination in the science of cybernetics.[37]

Herbert Marcuse was profoundly influenced by Heidegger's philosophy, as well as the theories of Marx and Freud. Marcuse was concerned with the extent that the dominance of the technological imperative is leading to the construction of an irrational society based on structures of domination and conformity rather than free relations developed through critical reasoning.[38] He argued that instrumental reason has dominated human thinking by repressing any aspirations, ideas, or values that cannot be submitted to technical rationality.[39] He argued that the human capacity for individuality, spontaneity, and critical reasoning are suppressed by the technological imperative's drive to pacify existence and create an advanced state of social cohesion and conformity. Marcuse defined technology in broad terms and considered the technological imperative to be a profound threat to human individuality and freedom. All aspects of human life, including culture, thought, leisure, labour, relations of production and consumption, the identification and satisfaction of individual and social needs, and the interactions between the public and the private realms are reduced to forms that can be integrated into the technological society in accordance with the constraints and demands of the technological imperative. Only those thoughts and activities that conform to these constraints and demands are included and empowered; opposing or alternative thoughts and actions are repressed or suppressed as being "irrational", "utopian", or "impractical". This threatens to construct a totalitarian society within which the operations of a system of domination pervades and mediates each and every thought and action.[40] In advanced industrial societies, the political realm is reduced to a technocratic system of institutional arrangements to control economic and societal development, where ideology is little more than a propaganda tool for legitimating this system. The technological imperative reduces every policy and legislation to a series of technical refinements and innovations of the technocratic system, which provides empowering mechanisms only to those who conform to it and further reinforce its power and legitimacy by doing so. Every qualitative social change is either reduced to a quantitative refinement of the system or remains impotent and imaginary. Technical rationality is necessarily a form of societal domination and control. Hence, as Marcuse argued, the totalitarianism of the technological society is not necessarily enforced by any tyrant or police state, but by the overwhelming and anonymous technological power and efficiency of that society. The use of coercion and police state tactics are only necessary when the technocratic system is incomplete and people still resist the technological imperative to construct the technological society. In industrial capitalist societies, individuals are increasingly integrated into an economic system that demands the total accommodation and submission of all human beings into that system by empowering those human beings who conform to the system and disempowering those who refuse to. This leads to a selection mechanism in favour of those individuals who conform to the expectations, norms, and practices of the relations of production and consumption. This reinforces the autonomy and legitimacy of the technocratic structures of planning and management, and eliminates any genuine possibility of radical social change or the development of alternatives. Once alternatives become impossible then society becomes "one-dimensional" and closes itself off from the possibility of the liberation of human beings and the rational development of the means to alleviate scarcity and need. Technical progress may well be defined in terms of the historical development and differentiation of means for achieving "the scientific conquest of Nature", understood as "the pacification of existence", but, rather than achieving "freedom from toil and domination" this has lead to "the scientific conquest of man".[41] Once human beings have conformed to the technological imperative, we become resources and objects within the technological framework available for administrative and managerial calculation, manipulation, and control; selected and ordered in accordance to with the latest techniques and technologies of administration and management. Our individuality is suppressed and we become alienated from our essence, our free and creative subjectivity, as beings capable of transforming our own existence. We cease to strive for our potential to live authentic and self-determined economic and social lives. We not only cease being able to transform the conditions of our existence, but, also, we cease to comprehend it and, therefore, the technological imperative leads to the suppression and repression of the potential for genuine social change and human emancipation.[42] The domination of human thinking by instrumental reason threatens to lead to the construction of a totalitarian technological society and the impossibility of human thinking, other than calculation and the expression of subjective preferences.

It was this concern with the construction of a totalitarian technological society that was central to Jacque Ellul's sociological critique of modern society.[43] He argued that *Technique* – the sum of all techniques – has impacted upon the development of the social organisation of human existence into a technological society. He argued that *Technique* is the totality of rational methods for the efficient organisation of human activity into a complex of standardised means for achieving predetermined ends, where the rationality of those ends is conditional upon the availability of standardised means to achieve them. In the technological society, *Technique* has dominated all human ideals, values, and aspirations, leading to a situation where efficiency and control have become ends in themselves. *Technique* has substantively pervaded all areas of human life and has superseded political ideology. Everything is judged according to its instrumental value for the further development of efficiency and control over the improvement of means. It transforms social institutions into an ensemble of technical solutions to social problems – a technocratic system – which is objective in the sense that it is transmitted like a physical thing through the rational organisation of relations of production and consumption. Western Civilisation has become a product of the autonomous development of the technological society to the extent that only that which is technical is considered to be part of civilisation. Anything non-technical is considered to be archaic, subjective, inefficient, or reduced to a technical form (e.g. consider how the study of fine art, poetry, or literature has become reduced to the analysis of techniques). The technological imperative is the perpetual search for the most efficient way to achieve any given task. It operates upon all facts, experiences, phenomena, and activities to reduce them to a set of techniques and their results in reference to technical standards, calculations, measurements, procedures, methods, and mechanistic causal accounts. Until it is replaced by a better technique, each reduction is represented as "the most efficient" technique available and the technical practitioner is obliged to use it to perform any given task in the most efficient manner. The technological imperative becomes a moral imperative. *Technique* is the sum total of all such applications of the technological imperative towards the discovery and use of "the most efficient" means to achieve any given task, and, therefore, it is the total reductive organisation of social being, in accordance with the dictates of the technological imperative. This tends towards the totalitarian organisation of humanity to maximise efficiency in every area of human endeavour. It is this totalitarianism that empowers and sustains the autonomy of technological innovation as the highest good – the perpetual search for the most efficient technique or device to achieve any given task. The goodness of the project of constructing the technological society – as a societal aspiration – is premised upon the assumption that human freedom and well-being are contingent upon the liberation from the limitations and vulnerabilities of our natural state. It is a grand experiment – a societal gamble – in the emancipatory potential of science and technology to liberate humanity from toil, suffering, and ignorance by constructing a paradise on Earth as a world of our own making. The construction of the technological society appropriates and absorbs the natural world, removing spontaneity and ambiguity, replacing it with reproducible and calculable relations of production, and, thereby, promising to create a more controllable, intelligible, and predictable world for human beings.[44] Ellul argued that, even though the experimental sciences of the seventeenth and eighteenth centuries did not provide the motivation for the application of the technological imperative to all areas of human endeavour, they prepared the way for the nineteenth century's positivistic conceptions of societal progress being represented in terms of the conditions for the construction of the technological society.[45] The application of mathematical rigor to the design of any machine was not only constrained by the demands of practical applicability, but preconditioned the prejudice that progress involved experimenting with the development of the designs most amenable to mathematical calculation and practical utility. While science provides general explanatory theories, it requires the application of technique as a necessary condition of its existence, and, since the nineteenth century, has become increasing the instrument of the rationalisation of the application of techniques to all areas of human activity.[46] In the twentieth century, science became dominated by the technological imperative to the extent that the development of atomic theory and the atomic bomb were necessarily connected.[47]

However, Ellul did not presume or advocate technological determinism. He maintained that the situation could always differ from the contingent actuality and rejected any notion of technological determinism. *Technique* can only become an autonomous "blind force" if we close our eyes to possible alternatives and repress our capacity to choose between them. It is our complicity in the technical rationalisation of all human activities that leads towards the reduction of all aspects of life to the technological imperative and the "inner logic" of maximising efficiency. As he advised, in the introduction to *The Technological Society*, it is helpful to think of freedom "dialectically and say that man is indeed determined, but that it is open to him to overcome necessity, and that act is freedom. Freedom is not static, but dynamic, not a vested interest, but a prize continually to be won".[48] Criticisms of Ellul's philosophy of technology as being overly pessimistic have simply ignored the spiritualism that he advocated as an alternative.[49] Furthermore, Ellul did not view modern society as some monolithic structure, constructed in accordance with some unitary essence, but involves tensions and struggles between competing tendencies and influences. The construction of the technological society is an inherent and dangerous tendency towards conformity to the technological imperative, but nothing have been determined and we can still win our freedom by choosing not to conform to the demands of the technological imperative. Indeed, for Ellul, the dominant tendency is towards the construction of the technological society, dominating the institutional arrangements of the State, transforming them into technocratic systems of public administration and regulation, but human individualism and a "laissez-faire spirit" tends to oppose the technological imperative and the dominance of technocratic rationality. However, we cannot rely on any allusions to the "free-market" and the "laissez-faire" pretensions of liberal capitalism, which Ellul considered to be a form of anarchism, because capitalism has become eclipsed by the efficiency of technical methods and the rationalisation of systems of production, distribution, advertising, and consumption.[50] Liberal capitalism has become transformed into industrial capitalism and even when it distorts the technological imperative, say by maintaining the use of inefficient technologies by suppressing patents or building in obsolescence, with all the concomitant waste and unsustainable consumptive patterns, this is only a corruption of how the technological imperative is implemented, rather than a genuine alternative or opposition to it.[51] The patterns of production and consumption generated by industrial capitalism (with its waste, pollution, and social costs) are inefficient and exacerbate the problems of scarcity, but they are still constrained and directed in accordance with the technological imperative and the construction of the technological society. As a result, Ellul interpreted the Marxist critique of the inefficiencies of capitalism and its contradictions with the technological imperative as being based on the presupposition that the technological society is essentially good and emancipatory, and, therefore, requires the liberation of the development of the technological society from the anachronisms of liberal capitalism and the corruption caused by industrial capitalism. Ellul considered the Marxist vision of communism to be one orientated towards technical progress, in all areas, giving free play to all technical automatism in every field of human activity, in order to achieve maximise efficiency, and, thereby, its fundamental result would be the construction of a totalitarian technological society.

The device paradigm

José Ortega y Gasset's *The Revolt of the Masses* has often been cited as a critique of direct democracy and a defence of liberal democracy, in favour of competitive elitism over majority rule.[52] However, this is something of a misrepresentation. It is actually a critique of "mass society" within which standards and directions are constrained in accordance with a conception of universality as being that which is common to all. It is a critique of the development of society in accordance with the anticipated needs of "everyman", which is, of course, no one. It is a critique of consumerism and conformity. Industrial society based on liberal democracy has become a technocracy and there is a widespread disregard for any critical reflection on epistemological and moral standards. The conditions for human individuality and excellence have been suppressed and eroded by "mass society" to provide the conditions suitable for "mass man". The "mass man" seeks only improved levels of consumption and security, avoids effort and struggle, and, for Ortega, is the product of societal inertia. Both "mass society" and "mass man" are the results of abundance and comfort produced through relations of mass production and consumption, the products of scientific and technological developments, rather than being the sum total of a multitude or the expression of some herd instinct. The "mass man" is an artificial construct, who is made possible by modern society, but refuses to take responsibility for societal development, insisting that this is the responsibility of the State, while asserting that, as "an individual", s/he has an absolute right to security, comfort, employment, income, and privacy, and it is the responsibility of the State to provide the conditions for these rights. "Mass society" is the ongoing reduction of liberal democracy into a technocratic system of regulated relations of production and patterns of consumption, wherein "mass man" is an entirely economic mode of being. In this respect, Ortega's conception of the "mass man" and "mass society" pre-empted Albert Borgmann's conception of "the device paradigm".

Borgmann proposed and developed this idea of "the device paradigm" to describe and analyse the way that the structures of modern society have stabilised into particular patterns of technological activity.[53] Borgmann defined modern technology, in a way that would be consistent with Heidegger and Ellul, as the "characteristic and constraining pattern of the entire fabric of our lives" in modern society.[54] His critique of modern society is based on the argument that "the device paradigm", in the form of stable patterns of production and consumption, erodes the traditions within which our practices gain their meaning. He acknowledged that science and technology have provided powerful devices and techniques to improve human health, agriculture, mobility, etc., which have overcome the limitations of our natural state of being and have provided the means to relieve the burdens of disease, hunger, and vulnerability, but he was deeply concerned with the way that technological activity has also eroded the meaning and character of human existence by eroding the conditions under which we discover meaning and character. Borgmann described these conditions in terms of a culture of focal practices and things that require our attention, care, effort, and patience, which place challenges and demands upon us and lead to the development of virtues, discipline, celebration, shared experiences, all of which he considered to be part of human well-being and excellence. By attending to focal practices and things (which resonates with Heidegger's understanding of "gathering"), we reawaken our connections with those practices and things in our everyday lives, focussing human relations in a way that engages us with each other and provides continuity within traditions. Focal practices and things are ends-in-themselves and are intimately bound-up with our social being and identity as human beings, gathering together particular experiences, human relations, and cultural meanings. By describing how they relate our everyday practices to traditions and also demand attention to detail, shared experience, commonality of purpose, and the development of virtue (understood in an Aristotelian sense), Borgmann described how focal practices and things have a "commanding presence" within a tradition that teaches us their meaning.[55] Examples of focal practices and things would include social activities, such as the traditional family meal, learning and playing music, conversations between friends or family, and ceremonies, but they can also include individual activities such as long distance running, walking in natural landscapes, and crafting tools or objects. Focal practices and things relate the particularities of our experiences to the generalities of traditions, unifying them through cultural meanings and providing connections between means and ends, individuals and their community, as well as enjoyment and celebration of them as ends-in-themselves. It is this lived relation between the particular and the general that gives our lives a sense of continuity, meaning, and value. He termed this as a "telling continuity" to describe the way that this lived relation situates the particularities of experience within the larger continuity of one's life, community, history, and a sense of place in the world. It is through the act of focussing upon these practices and their meaning that the focal thing "gathers the relations of its context and radiates into its surroundings and informs them".[56] Focal practices and things are the specific practices and things of our lives that are of great importance for our sense of well-being and our reflective care for the good life. Hence, for Borgmann,

> A focal practice, generally, is the resolute and singular dedication to a focal thing. It sponsors discipline and skill which are exercised in a unity of achievement and enjoyment of mind, body, and the world, of myself and others, and in social union.[57]

Even though technology ties us to an active and engaged relationship with the world through labour, when that relationship becomes a commodity, understood only in terms of exchange and consumption, we lose our capacity for a deeper, bodily, mental, and social relation with the world. The technological society – driven by the capitalist economic imperative to maximise profits – has trivialised life and turned human beings into passive and isolated commodities and consumers. The democratic and rational revaluation of technological innovation must be based upon a community engagement with the articulation and critical assessment of the concrete proposals of technology in contrast to the current conditions of their fullness, in accordance with considerations of sufficiency for a practical appraisal of the good life.[58] While modern technology has evidently increased human power by providing an ensemble of devices and techniques, it has also reduced our commitment to developing our social relations and human excellence by eroding our need for focal practices and things. How has modern technology eroded our need for focal practices and things? For Borgmann, modern technology is the ensemble of disposable devices and techniques designed to alleviate effort and glamorous in their appeal. They are disposable in the sense that, situated in everyday life as mere means to ends, they are replaceable and, thereby, have only an instrumental relation to the ends to which they are directed towards satisfying. Any device or technique may well be thrilling or glamorous, in terms of the powers or opportunities it affords, but it also can be replaced with a new, improved device or technique and, therefore, it does not have any intrinsic value or meaning. While devices such as the computer or the telephone may well allow us to be remotely connected to distant people and places, they remain mere means to an end as a medium for communication. The human relation with devices and techniques is either one of invention or use; an economic relation of either production or consumption. Devices, such as the motor car, for example, are mere means to ends, which are exchanged, via the medium of money, as instruments to satisfy our present needs (travel to work, take the kids to school, attract members of the opposite sex, attract the envy of members of the same sex, etc.) as well as afford us future opportunities. Even though we may well be seduced by the shape, colour, and power, motor cars remain instrumental, given that they can always be replaced with a better or the latest model, or an alternative means of transportation. Of course motor cars can be transformed into focal things, such as vintage cars, lovingly restored, carefully maintained, driven around town or in rallies, as a pleasure for its own sake, and understood in terms of the history of its design, development, and usage, as well as made meaningful through personal experiences. Whether a motor car is a focal thing, an end-in-itself, or a device, a means to an end, is determined through our relationship with it, rather than any essential category of how it came into being (i.e. through advanced industrial engineering and relations of production) or its intrinsic properties. It is possible for a mass produced motor car to become a focal thing, just as it is possible for a hand made and restored vintage car to remain a device. As a focal thing, a particular motor car has inherent value as an end-in-itself, a work of art, made meaningful in terms of everyday focal practices. As a device, motor cars are used as mere means until they are discarded as junk, instrumentally valuable for parts or recyclable materials. While it is evident that people do develop fetishistic relations with devices, such as motor cars (or televisions, or music systems, etc.), these devices remain disposable as things in the sense that the relationship is with the glamour of the device, the sense of ownership of the latest model, and also the functionality (or prestige) it affords. It is possible for the same vintage car to be a focal thing or a fetish depending on how human beings relate to it.

Borgmann argued that devices replace focal practices – the microwavable meal replaces the traditional cooked family meal, the television replaces conversation and other entertainment, and the electronic music system replaces live performances by musicians – and transforms our social relations with the material conditions of our existence and how we interpret their meaning and value. Devices erode our traditions by making them obsolete or quaint (parochial), which replaces the commanding presence and telling continuity of focal practices with the thrills and fetish of using the latest disposable device to achieve culturally given ends, which may well not be achievable without such devices. This pattern of production and consumption erodes our need for focal practices and things because "the peril of technology lies not in this or that manifestation, but in the pervasiveness and consistency of its pattern".[59] It was for this reason that Borgmann acknowledged the *affluence* of the device paradigm and distinguished it from the *wealth* of engagement through focal practices and things, and called for restraint of the device paradigm in favour of a return to focal practices and things as ends-in themselves.[60] Borgmann was critical of Ellul for being overly pessimistic and deterministic about modern technology, and, following Heidegger, he was also critical of instrumentalist and anthropological philosophies of technology.[61] While his idea of focal practices and things was clearly inspired by Heidegger, it was also inspired by the writings of Emerson, Thoreau, Melville, and Aldo Leopold, as well as Native American oral traditions.[62] Deeply rooted in this American narrative tradition, Borgmann was concerned with how modern technology – as *Gestell* – has concealed Being within a framework of instrumentality for future use, and, he was also concerned with the extent that we have complacently become quite unthinking about technology, which has in many respects become an ensemble of unconsidered means to achieve ill considered ends. "The promise of technology", as Borgmann termed it, is the two-fold promise of liberating us from the ills and limitations of the natural world and, also, embracing a vision of the good life that technology offers to make possible.[63] Hence, Borgmann was aware of the societal gamble and the implicit faith in science and technology as providing the potential to construct a better, more certain, and rational artificial world, the technological society, as a replacement for the natural world. The societal gamble is the faith emergent from the Enlightenment tradition that science and technology will liberate us from superstition, fear, toil, suffering, and perhaps even death, allowing human beings the freedom for creativity and to become masters of our own destiny. However, we have strayed considerably from this vision. We have become enslaved within patterns of production and consumption. The Enlightenment vision of a rational society populated by free and equal human beings has become reduced to a positivistic measure of our "standard of living" calculated in relation to our level of consumption. A world of intrinsic value has been replaced by a system of devices, commodities, and exchange-values. The enjoyment of a meaningful present has been replaced with the thrill of purchasing and using the latest devices as the measure of well-being. The increasing complexity of technology is concealed using "user friendly" push-button or point-and-click interfaces, which increases the separation between production and consumption, further disconnecting means and ends. The internal construction of the device (e.g. a mobile phone) is concealed completely within its casing, as a set of functions to deliver a commodity or service (convenient communication). This not only increases the disposability of the device, but it increases the dependency of the consumer upon the producer. Despite the increased complexity of technology, devices deskill their users by alleviating the burden of having to learn how the device works, ironically using modern technology to disconnect the consumer from "the technical" and allowing the consumer "to use up an isolated entity, without preparation, resonance, or consequence".[64] Technology is inbuilt into society, into societal structures that constitute "the inconspicuous pattern by which we normally orientate ourselves".[65] The device paradigm – the overall pattern of use of the ensemble of devices – transforms the natural world in standing-reserve, structured in its instrumentality within patterns of production and consumption. Yet, asserted Borgmann, increasing levels of production and consumption does not lead to satisfaction, despite the promise of technology to bring liberation and enrichment. We are left alienated from deeper cultural and personal meanings, disengaged from our deeper human aspirations, or simply mindlessly locked into patterns of consumption. Borgmann argued that we need to provide genuine alternatives to this unreflective consumptive relation with technology by reawakening focal practices and things, once again becoming mindful and respectful of their commanding presences and telling continuity. According to Borgmann, we need to take time to develop interpersonal relations with each other through these reawakened focal practices and things, paying close attention to how these practices and things are meaningful within our individual and community lives, aiding the development of our attention, skills, capacity for effort, carefulness, and understanding of our history and traditions. This requires a commitment to our everyday activities as being ends-in-themselves. It requires a return to the enjoyment of a shared present, for example, over a shared cooked meal at the family table, or while sitting around a wood burning stove, or during a conversation, or while walking along cliffs, through woods, or across hills, or any of the everyday activities shared with family, friends, or neighbours. It is only through the collective devotion to the meaning of our lives that we will be able to recover these meanings, reawakening the possibility of human excellence and the good life through the effort required to learn and develop reflective and mindful attitudes to our lives.

Technology acts as a substitute for other activities. For example, television acts as a substitute for conversation or public entertainment (such as going to a cinema, a theatre, or a concert) and, in many respects, "the device paradigm" has distracted us from "great embodiments of meaning", as Borgmann put it.[66] However, as I argued in *Modern Science and the Capriciousness of Nature*, the erosion of traditions and cultural meanings has not been caused by the development of modern technology in society, but, instead, technology acts as a substitute for traditional practices and, thereby, conceals the erosion that has already taken place.[67] Otherwise it would be difficult to explain why focal practices and things, with their "commanding presence", would be so readily eroded by the device paradigm. We would have to explain why human beings choose the thrilling over the meaningful if, as Borgmann claimed, the meaningful is an essential aspect of human identity. Of course one can appeal to the convenience and ease of using modern devices in relation to the difficulties involved in learning traditional practices, but, in which case, one would also need to explain how focal practices and things lost their "commanding presence". If one only learns the intrinsic value of focal practices and things through the attention, care, and effort of learning traditional practices, through focussing, this still does not explain why human beings stopped valuing them as goods or ends-in-themselves. In my view, it is only when sitting around the hearth, to use Borgmann's example, had already lost its value as a focal practice, that the transition to electric heating, delivering heat as a commodity, became desirable and the focal thing, the wood burning stove, lost its "commanding presence". Otherwise Borgmann's argument would entail that human beings began considering wood burning stoves in terms of their instrumental value, as a device to deliver warmth, but subsequently learnt their intrinsic value from the pleasure and meaning that their preparation and sitting around them brings, but if this is the case, then there is no reason to presuppose that new focal practices and things will not emerge from sitting around in an electrically heated room. Borgmann's rejection of this possibility seems to draw something of an arbitrary line between kinds of technology based simply on his subjective estimation of the sufficiency of any technology. Hence, according to Borgmann, playing a musical instrument is a focal practice but playing a DVD is an act of consumption, even though in both cases the focus would be a group of people coming together over the shared enjoyment of the music. After all, could we not equally argue that the wood burning stove eroded the communal gathering of a village or tribe around a shared fire, it secrets only known by revered elders? Did not the wood burning stove erode the magical and communal ceremony of fire-making? Did not the tinder-box erode the "commanding presence" of making fire using dried twigs and moss? Taking this argument to its Promethean extreme, one might even argue that the knowledge of how to make and use fire eroded the "commanding presence" of the natural world and the traditional practices our ancestors had developed to survive in it.

It seems from Borgmann's argument that his objection is to the effortlessness of modern technology as a means to satisfy our desires. Hence the focal thing is not so much the wilderness experience while backpacking or the splendid trout caught by fly fishing, to use his examples, but it is the effort involved in having those experiences. He considered effort to be a good because it because of its value in building character through making particular kinds of experience possible through effort. The problem is not modern technology at all, but is our indifference to the development of our own characters, if, like Borgmann, we consider the development of good character to be an intrinsic good. But, what does this mean? By unburdening us from making the effort necessary to achieve any given ends, by providing the end as a commodity to be purchased, the device disengages us from caring about the conditions of our existence. Our existence becomes mediated by "the device paradigm" and is dependent upon it; the quality of that existence is to be purchased through participating in an economic system of exchanges. If this is the case then the problem of "the device paradigm" is that tragically its success in satisfying our material needs has eroded the conditions under which we develop a meaningful and existential relationship with the conditions of our existence. Conformity to "the device paradigm" results in such an easy life that we are not compelled to engage in the practices that develop our characters as beings embodied in a physical world. We become shallow consumers, without any regard for understanding the conditions of our existence. However, my argument is that the problem of technology is not that it caused us to become shallow consumers, but that it conceals from us the extent that we have become shallow by removing the condition under which we would confront that fact; by supplying the means to satisfy our needs, it conceals the causes for the erosion of our desire to maintain the traditional practices which previously satisfied our needs. For example, television and microwave meals are not responsible for the demise of the traditional cooked family meal, but, by acting as substitute for this traditional practice, the TV dinner, by providing an alternative means to prepare and eat food while being entertained, has concealed the erosion of traditional family relations that have already occurred. What modern technology does is conceal from us the reasons why our traditional practices became unsatisfactory in the first instance by providing us with an alternative that makes those traditional practices obsolete as a practical activity. Once we take this into account, we can see that underwriting Borgmann's call for a return to traditional focal practices and things as the means to develop character is a set of moral prescriptions and norms for social being. This somewhat conveniently neglects the inequalities that were involved in many of these focal practices. For example, it normalises the gender based division of labour that is involved in the traditional cooked family meal and by placing the blame for its demise on modern technology it has ignored the social changes that have taken place where women no longer need to passively accept their traditional role as unpaid housekeepers and cooks. What modern technology did was simply compensate for this social change by allowing the family to eat without depending on women to cook and serve the meals, while also continuing to shield men from the burden of learning how to cook. It would be quite impossible to return to this traditional practice without re-imposing the gender inequality inherent to the traditional division of labour within the family, but what modern technology has done is prevent a critical examination of the reasons why men are traditionally reluctant to take a share in the preparation and serving of meals by making the question obsolete. However, the traditional attitudes and prejudices have remained intact. Once we acknowledge this inequality, then we should be aware that recovering the focal practice of the family meal should involve consciously, mindfully transforming it to accommodate social changes by equitably sharing the burden of learning how to cook and prepare meals throughout the whole family, regardless of gender. This would not be a return to traditional practices at all, but would be a focal accommodation to the fundamental changes that have occurred within the family structure, while recovering those focal practices from "the device paradigm" that concealed the need for such an accommodation.

Putting aside the essentialism and romanticism inherent to Borgmann's philosophy of technology, as well as the normative moral conservatism which underwrites his choice of focal practices and things, my criticism of his argument is the technological determinism which underwrites his causal connection between "the device paradigm" and the erosion of traditional focal practices and things. In my view, his argument is akin to claiming that the bullfighter's red cape caused the death of the bull because it distracted him from the bullfighter and his sword. My argument is that the pernicious aspect of technology is not that it causes mindless and unsatisfying lives, as Borgmann claimed, but, much worse than this, it conceals the irrationality and arbitrariness of the relations and structures of modern society by providing a successful system of providing means to distribute commodities for those who conform to the system and are able to integrate themselves within it. It does not undermine a philosophical and spiritual engagement in life, but conceals the fact that such an engagement is absent. For Borgmann, the meaning of rationality was bound together with a deep philosophical, moral, and creative vision of the good life. In his view, we can consider ourselves to be rational to the extent that we are capable to reflect upon and understand the conditions of our existence, as well as possess the skills and means to control or change them, in relation to moral norms and ideals for human excellence of character. His argument presupposes that "the device paradigm" does in fact, in itself, prevent us from being rational by disengaging us from the material and moral conditions under which we develop character through both the physical and intellectual effort required to envision and realise human well-being. Even if we agree with Borgmann's claim that modern society lacks an explicit vision of the good life over and above a certain level of consumption, we would still need to show that "the device paradigm" has also undermined such a vision. My position is that it is not evident that "the device paradigm" has done this, but, due to its success in providing commodities and satisfying material needs, it has hidden the need for such a vision from public awareness and deliberation. By covering our irrationality and arbitrariness, with technical rationality, we have become unconscious of our irrationality and arbitrariness and no longer confront them. The pervasive conformity underwriting the development of the structures and content of society, constructed without any reflection upon their reason or consequences beyond their immediate instrumentality within the system, remains unchallenged. Moreover, even if it is true that people in modern society are in large part significantly unhappier than people in pre-technological or traditional societies, as Borgmann presupposed, we still have to show that it was the development of "the device paradigm" which decreased human happiness. Putting aside the considerable difficulties involved in evaluating the levels of happiness or unhappiness of people in different societies, cultures, and historical eras, it seems evident that, even if we could make such a comparison, we would be still confronted with the difficult (perhaps impossible) task of demonstrating the existence of a causal relation in order to argue that "the device paradigm" was the dominant causal factor responsible for the unhappiness of people within modern society. While it is not at all evident that people in modern society are generally unhappier than people in traditional societies, even if it is true that they are, it is arguable that "the device paradigm" did not cause their unhappiness but distracted them from the true causes of their unhappiness by allowing them to satisfy their material wants and needs without having to discover and confront the sources of their unhappiness. Rather than blame technology for all our ills, putting our faith in a somewhat reactionary call for a return to traditional practices, we need to carefully examine why our traditional practices eroded in the first place. In my view, the erosion of our traditions occurred because they were inadequate or inappropriate due to their inherent incoherence, inequalities, and inconsistencies, and, as a result, they could not withstand societal changes. Rather than put our faith in tradition, we need to critically examine our personal and societal goals alongside understanding whether and how societal structures empower, distort, or suppress them. In some respects this will involve critically reflecting on how "the device paradigm" has failed to live up to the promise of technology – how our faith in the societal gamble is misplaced – and in other respects it will involve critically reflecting on how our traditions failed to realise their ideals and values, either by failing to live up to them or by being riddled with internal contradictions and falsehoods. If there is some truth to my argument, then the increased development of "the device paradigm" will not make us unhappier, but our conformity to it will further distract us from confronting and changing the causes of our unhappiness by continually substituting the thrills and glamour of new devices and higher levels of consumption for the need for philosophical reflection and critique of our vision and choices regarding the development of society. "The device paradigm" does not suppress our capacity for reflection and critique, but simply distracts us from the conscious recognition of their necessity to prevent the irrational development of the technological society. Without such reflection and critique, the technological development of society substantively structures and conditions the possibilities and directions of our modes of engagement with the conditions of our existence by increasingly systematising human relations into systems of media and exchange. This will further superimpose technical systems over our intentions, expectations, limits, and interactions, until it has reduced them to those capable of being satisfied by technical systems. At which point all human relations will be reduced to "the device paradigm" and Sophocles' warnings about the tragedy of the human condition will become a nightmarish reality, wherein, like all but one of the denizens of Aldous Huxley's *Brave New World*, our descendants will simply not care for anything other than their functionality within the system.

3
Democratising the Technological Society

Technology is an historical agent for change by changing the material conditions of human existence and extending our horizon of possibilities – it is part of the cultural stock of knowledge, devices, and resources which transforms social, economic, and political possibilities and overcomes previous limits. Technological innovation has a substantive impact on the structure and content of the development of society. The ongoing proliferation of inventions and techniques conditions human choices by reducing the difficulty of realising them and also by making new choices possible. It is quite uncontentious to claim that human existence – certainly as we know it – is dependent upon a complex, interconnected network of technological infrastructure comprised of systems, devices, and techniques. The continuation of specific forms of technology is a condition for the continuation of specific forms of social order. Technological innovations have far reaching and unforeseeable social consequences and our dependency upon technology is irreversible, dialectical, and deeply psychological. However, does it follow from these facts of our existence that the development of technology has its own "inner logic" or laws? – A life of its own, so to speak? Is technology an autonomous agent for change? Are the trajectories of technological innovation the consequence of historical necessity or human choice? The argument that I shall present below is that the trajectories of technological innovation are dialectical in the sense that they are shaped by human choices and, in turn, technology shapes human choices. The trajectories of technological innovation are conditioned and contingent upon the current state of scientific research and technological development, how we understand the world, our nature, and how we represent the alethic modalities of human existence (our possibilities, necessities, and our limitations), all of which substantively condition our expectations, intentions, judgements, and imagination. Historical necessity is itself an historically conditioned and contingent product of human choices, made possible by new technologies, and these developments entail substantive demands upon human beings to use those technologies and develop them further. For example, the development of medical technologies has been contingent upon human choices to develop such technologies, but the opportunities afforded by them to improve human health and prolong human life, given the human desire to be healthy and to live longer, entails substantive demands to use these technologies, as well as develop them further, as a moral imperative, while also bringing with them a host of social problems (such as the economic problems associated with the demographic ageing of the population), environmental problems (such as those associated with high populations), moral dilemmas (such as those associated with birth control, abortion, cloning, organ transplants, artificial insemination, genetic screening, euthanasia, etc.), as well as political problems (such as those involved with the funding of medical research, protecting intellectual property rights, and the equity of access to increasingly expensive healthcare, etc). The apparent autonomy of technology is a consequence of its abstraction from the cultural structures and relations within which it is tacitly and implicitly embodied and given meaning, purpose, and functionality. Once we recognise and explicate these cultural structures and relations then we are better placed to understand how technology changes them, which improves our understanding of how we can choose to develop and differentiate technological innovation along alternative trajectories. However, the unquestioning acceptance of the societal gamble (further entrenched in conformity to the device paradigm) leads to a cultural repression of this capacity for choice, resulting in technological determinism at one extreme and the positivistic representation of technology as the value-neutral patient of human decision at the other (what Heidegger termed as the anthropological and instrumentalist positions). Both of these extremes involve maintaining an unthinking and uncritical relation with technology, which allows the technological imperative to be represented as the rational response to historical necessity based on the current state of our objective scientific knowledge. Whereas, if we critically examine the historically conditioned and contingent trajectory of human choices that have culminated in this representation of the technological imperative, then we can understand the societal gamble as being the human, cultural faith in the possibility of creating the kind of society that will technologically empower human beings to rationally develop a post-scarcity society within which human beings are liberated from historical necessity. In other words, the technological imperative is the consequence of the societal effort to establish ontological (realist) or epistemological (positivist) representations of technology as the sole agent for historical change and the creation of the conditions for human rationality and freedom.

In his recent writings, Heilbroner has further softened his "soft" technological determinism.[1] He now proposes that his position is a heuristic for the investigation, analysis, and interpretation of historical socioeconomic events in terms of the directions and limits of technological innovation and development. It is to provide a framework of explication, rather than historical laws or predictive models, given in terms of the patterns and consequences of technological development. Its purpose is to help to develop an awareness of the substantive impact of technological change upon societal development, while also remaining aware that human choices and cultural attitudes effect technological development. Heilbroner argued that "soft" technological determinism should be epistemologically judged on its psychological value for aiding us to write an intelligible history between the extremes of a fully determined or undetermined state of affairs. His heuristic helps us to understand the technological preconditions of each stratum of technological development and how these effected the subsequent changes in social organisation and the directions of further development, but it does not presuppose that these developments were inevitable once their preconditions were met; nor does it presuppose that these developments followed a predictable evolutionary trajectory. We can understand how the steam mill required the development of the hand mill, along with the societal changes that followed the invention and dissemination of the steam mill, but once we reject the idea that the steam mill was inevitable once the hand mill had been developed then our theory significantly departs from technological determinism in the sense of historical determinism, and, moreover, once we recognise the historical conditions and contingencies involved in the technological development of the sciences of mechanics and thermodynamics then our theory departs from nomological determinism as well. By postulating that the trajectories of technological development are emergent and refined within an historically developed technological framework, itself set up to satisfy human ambitions and expectations, which is incomplete and experimental, we can explain the sequencing of the stratification of technological development in terms of increased complexity, rather than appealing to either natural or historical laws. The "inner logic" of technological convergence can be understood in terms of the structures of the technological framework and their interconnection within wider society. Bimber argued that Heilbroner's softening of his technological determinism has sufficiently retreated from nomological determinism to disqualify his theory from being a kind of technological determinism at all. Instead, Bimber argued that Heilbroner's theory is a normative theory, within which technological development remains an important influence on the historical development of societies, but it is not the only influence, given that cultural differences in societal goals lead to differences in technological developments, within the objective limits and possibilities allowed by natural law. In my view, Heilbroner has considerably moved away from technological determinism towards a substantive and critical theory of technology, such as Andrew Feenberg's critical theory of technology.[2]

Following sociological and historical accounts of the conditions and contingency of technological design, implementation, and development, Feenberg has developed a substantive and critical theory of technology as a critical response to technological determinism.[3] For Feenberg, technology does not have any essence or "inner logic" at all – there is no such thing as technology "in itself" – but exists only in the contexts of its employment and can be adequately understood only in terms of how it is used in context. Hence, according to Feenberg, there is an ambivalence of means, allowing new employments of technologies, which changes the kind of industrial society or civilisation that is actually constructed, including its political institutional arrangements and relations of production. He argued that technology does not require or determine a particular form of society, but provides a potential for various divergent directions of cultural development, which he terms as "civilizational projects". His analysis of technology utilised a distinction between "primary and secondary instrumentalizations". "Primary instrumentalization" characterised the human attitude, orientation, or relation with the natural world or reality. According to Feenberg, it was this aspect of technology dominated the substantive theories of technology developed by Heidegger, Marcuse, and Ellul. He was critical of their neglect of the actual contingencies of technological development, the choice of particular technologies in response to particular decisions about societal development, which he termed as "secondary instrumentalization", and argued that a substantive and critical theory of technology must analyse technology in terms of both these aspects, if it is to avoid essentialism or determinism. By understanding "secondary instrumentalization", Feenberg hoped to develop a theory of technology that would show how a dialectics of technology, properly understood in relation to practical activities rather than some "mysterious new concept of reason", could help to holistically reintegrate technological development with social development by focussing on the social contexts of technological activity and technical decisions. Thus ecological, aesthetic, and ethical aspects of social development would be integral aspects of the technological development of society. Human health and well-being would be integral engineering objectives, rather than social externalities or impositions. Feenberg argued that this can be achieved by recovering technical pluralism, democratising the technical sphere, and adapting technology to the requirements of a socialist society, as democratically decided in relation to concerns with the concrete outcomes of technological development. In this way, technological innovations could be designed to be incorporated into complex ecological and social environments, addressing contemporary social and environmental problems. This does not require centralised planning, nor does it leave it to the caprice of profiteering, but requires the democratisation of the technological society.

"The technical" as a subcategory of "the social"

Feenberg argued that historical and sociological accounts show that "the technical" is a subcategory of "the social". He considered technological determinism to be essentialist because it is based on the metaphysical assumption that the essence of "the technical" is defined by its "inner logic" of the rational application of the objective knowledge of natural laws and mechanisms, which are supposed to exist independently of "the social", and, thereby, the rational development of "the technical" requires expertise in this "inner logic". According to Feenberg, such essentialist claims lack any empirical or historical evidence, whereas historical and sociological studies show that the structure and the content of "the technical" are socially constructed in the sense that they are given meaning through intentionality, communication, evaluation, representation, and usage. There is rarely one best way to perform any given task and the decisions about how best to proceed are conditioned by and contingent upon establishing consensus and commitment between social groups, such as political, bureaucratic, technical, and economic agents, institutions, and organisations, as well as scientists and technicians.[4] Technical pluralism reveals the centrality of human choice to technological development.[5] Technical rationality is bounded by its context and, when we examine "the technical" as a subcategory of "the social", we can see how the technical transforms human identity in terms of technical agency – he termed this as "vocation" – and it is this transformation, including all of its ethical and aesthetic mediations, which has a substantive affect on human character, our moral landscape, and our future expectations.[6] Technologies are the products of a social history of interest-laden agendas inscribed in their form, structure, content, and trajectories of development, implementation, and usage, rather than any "inner logic". For example, the contemporary, public debates and dilemmas about biotechnology, energy production, military proliferation and research, as well as the debate about their environmental, political, economic, and social consequences, are the historical results of socially conditioned and contingent decisions that reflect the interests and distribution of wealth and power in the development of society and the means of production. The choices between alternative directions of technological innovation are not limited to criteria of technical or economic efficiency, nor are they based only on the scientific understanding of the intrinsic properties of technological systems and devices, but, they are made in accordance with the socially established coherence between technologies and the interests of the various social groups that influence and fund the research and development processes. In this sense, all technical knowledge, judgements, and artefacts are socially constructed products of human interests and power relations shaped by tacit preconceptions and theoretical representations about societal possibilities and objectives, providing norms for how society ought to develop its technological capacity in order to realise its possibilities and achieve its objectives. Human interests and power relations place limits and trajectories of further development of "the technical" and, once these directions have been chosen and developed in practice, after being integrated and embedded into the infrastructure of functionally interdependent technologies, it becomes increasingly difficult to explore alternatives because these technologies have become an everyday, ordinary part of practical activity. Technologies are both agents and patients of societal development of economic and political resources and institutional arrangements bound up with practicality, understood as the reproduction of everyday, ordinary life experiences and activities. It is this everyday ordinariness which resists the development of alternatives when any change in "the technical" involves changes in "the social" which initially will be impractical or threatening. In this way, human interests and power relations become embodied in societal infrastructure, machines, architecture, bureaucratic administrations, and institutional arrangements, as well as the directions of scientific research and development.

Feenberg termed these embodiments as "technical codes", which represent a structural inertia that empowers specific directions of technological development, as a set of technical habits and tendencies, while acting as a resistance to alternatives by raising the stakes and costs of trying alternatives. Technical codes trace the structural reproduction of the power and authority of those with privileged access to and control over "the technical". In turn, "the technical" imposes demands upon the privileged class, if they are to maintain their position of privilege and, consequently, technical codes express the rules of the structural reproduction and development of the norms of a class-based society. Technical expertise is the knowledge of these technical codes rather than the understanding of any "inner logic" of "the technical". In this respect, technical codes are abstract indexes and rules for ordering technical activities and they reveal how "the technical" has become tightly bound within networks of social power, as a system of productive and destructive forces and relations, structured into political, economic, and military institutions and organisations. Expertise is a conscious attunement to channelled social power rather than a possession of objective knowledge, even if technical codes are represented in terms of objective structures and mechanisms.[7] Hence, following Feenberg's argument, "the technical" can be understood in terms of the ongoing, constructive and destructive interactions between organised, channelled economic and political resources and institutions that influence and limit the trajectories of research and development, in order to reproduce and enhance class interests, structuring the technological provision of new societal possibilities and aspirations, as well as the resources and institutional arrangements that the employment of "the technical" makes possible. Typically scientific research and technological innovation are implemented, as David Noble put it,

> ...only after the decisions have been made to invest social surplus in their development and use, and these decisions were based not only upon mere guesses as to their technical and economic potential but also upon the political interests, enthusiastic expectations, and culturally sanctioned compulsions of those few with the power to make them.[8]

Technological determinism is a "decontextualisation of technology", as Feenberg termed it, because it ignores the social shaping of technology.

This aspect of his critique of technological determinism is important because it highlights the political value of specific technical codes for the purpose of being able to publicly establish particular technical decisions as being the rational, efficient choice, and, therefore, once we take the social conditions and contingency of technical codes into account, we can critically evaluate the claim that the current trajectories and content of the technological innovation are the only rational choice available to us. They are dependent upon a "civilizational project". The philosophies of technology that justify technological determinism, such as positivism or realism, affirm or presuppose the rationality or historical necessity of the current trajectories of technological development and the social order they sustain. However, in my view, Feenberg did not properly take this into account because he assumed that his sociological theory of technology is value-neutral and therefore he asserted that any claims that technical rationality is objective are simply mistaken. Feenberg correctly pointed out that positivist and realist philosophies of science and technology entail moral assumptions, values, and visions of the ideal society, but what he failed to recognise is that these philosophies are premised upon a societal gamble in the goodness of the technological society as a "civilizational project". The proponents of technological determinism are not simply mistaken, as Feenberg claims, but are supportive of the current trajectories of technological development as a normative affirmation of modernity premised on the assumption of the goodness of the norms of technical rationality. Feenberg made the error of assuming that the idea of value-neutral technological development is false, a myth, or an illusion, when it is actually an expression of a commitment to a societal experiment. If this is true then technological determinism is not simply a species of "Whig history", as Feenberg put it, but is itself a partisan species of rhetoric directed to legitimate the *status quo*.[9] Technological determinism is the product of the intellectual suppression of alternatives, justified by representing the current trajectories of technological development as the most efficient means available, in order to sustain the social order of hierarchical structures of decision-making and the social inequalities it aims to preserve. Technological determinism is bound up with political ideology and the reproduction of propaganda, irrationally disseminated in order to legitimise the representation of the *status quo* as being either historically necessary or the only rational possibility. It is a variant of "the last man standing" argument. Rather than simply deny the truth status of technological determinism, we need to show that the accounts of the "inner logic" of technological development, often made through appeals to scientific knowledge of natural laws and mechanisms, entail values, moral judgements, and metaphysical precepts, and, by revealing these assumptions, we can show how the technological society, as a "civilizational project", distorts and suppresses human potential. A successful substantive and critical theory of technology will depend on the articulation of an alternative "civilizational project" that promises to recover and realise human potential, leading to a rational, egalitarian, and libertarian society. What we need to do, through hermeneutic reflection on the historical conditions and contingencies of the current trajectories of technological development, is to recover this alternative vision of society and subject it to reasoned deliberation and critical reflection, as a direct challenge to the hierarchical *status quo*. It is insufficient to show that historical and sociological accounts of the technological society reveal it to be conditioned and contingent, as if the obstacle to choice is our false consciousness about the human condition, when we do not actually participate in the processes by which choices regarding societal development are made and justified. In order for democratic participation to be genuinely possible it is necessary (but not sufficient) that the public are conscious of our potential for choice, but we also need the social power to gain access and control over the political processes within which deliberations and decisions are made. For genuinely democratic participation to be possible, we need a radical alternative to the current framework of deliberation and decision-making, and the social power to implement and develop this framework. Otherwise, we are none the better off by knowing that technological development is contingent, apart from being aware of how undemocratic our society is and how cheated we most of us are from achieving our full potential as social beings.

Feenberg used the example of the bicycle to point out that the final design of the bicycle that we take for granted today was the result of a long, historical struggle between different social groups about the purpose of a bicycle, rather than a derivative product of an abstract technical rationality.[10] While Feenberg is quite correct, he fails to acknowledge that a central feature of the bicycle is that it has two, circular wheels, and that there is a distinct, technical limit on the possibilities for any alternative design for a bicycle. Of course, it is possible to make a clown bicycle, with square or ovoid wheels of different size, etc., but the reason why such a bicycle is funny is because it breaks with the "technical limit" of the design of an efficient and stable machine performance of a bicycle. Its humorousness is in direct proportion to the inefficiency and instability of its performance. While the decision to design and build bicycles as a mode of transportation is socially constructed, the technical limits to the form of a bicycle are discovered, rather than decided. To use another of Feenberg's favourite examples: it is clearly the case that human choices were involved in dealing with the problem of exploding boilers of nineteenth century American steamboats.

> What a boiler "is" was [socially] defined through a long process of political struggle culminating finally in uniform codes issued by the American Society of Mechanical Engineers... Quite down-to-earth technical parameters such as the choice and processing of materials are *socially* specified by the code. The illusion of technical necessity arises from the fact that the code is thus literally "cast in iron" (at least in the case of boilers)... A properly made technical object simply *must* meet these standards to be recognised as such.[11]

The struggle between the politicians seeking votes, the boat owners seeking profits, the passengers seeking a means of safe travel, and crew seeking wages (and to live long enough to spend them), shows that what constitutes "a safe boiler" design is bounded by socially defined situations and solutions. This is an uncontentious point. The design of a safe boiler is evidently socially contingent in so far that not only is every proposed design a product of social activity, but the final choice is made in accordance with socially agreed criteria. It is also evident that there are historical and social conditions and contingencies at play in the creation of the social need for boilers and steamboats in the first place. However, in my view, Feenberg's critique of technological determinism overly emphasises the historical and social contingency of technical codes, and, thereby, gives an asymmetrical and reductive account of them as socially constructed objects. After all, it would seem reasonable to argue that two of the minimal conditions for the design of a safe boiler are that it turns water into steam and does not explode. This does not mean that a boiler that produces steam and has not exploded is "a safe boiler", given that it might stop producing steam or explode in the future, but it does mean that if it does not produce steam then it is not a working boiler, and, any boiler that explodes is not a safe boiler. The definition of the minimal conditions of the design of any machine is socially contingent, given that it requires social activity to clarify and communicate this definition, but whether the object of that definition satisfies it cannot be determined only by social activity alone. It is the case that "the technical" is a subcategory of "the social", in the sense that technical activity is a kind of social activity, but the consequences and limits of any particular technological activities are socially discovered, rather than decided. In my view, Feenberg's reliance on social construction theory ignores how social responses to "unintended consequences" – discoveries and accidents – also shape technological development, alongside human decisions about how to organise and develop technological activity to achieve human aspirations. But, how can we account for "unintended consequences" and "technical limits" without appealing to some kind of "soft" technological determinism, such as Heilbroner's, in order to account for the fact that sometimes boilers explode and bicycles need to have two circular wheels? How can we explain "unintended consequences" or "technical limits" without conceding that natural mechanisms and laws condition the content of technical codes?

The dialectics of technology

Indeed the trajectories of technological innovation are contingent upon choices that resist and suppress alternative paths of development, but these trajectories are also the result of discovered powers, technical limits, and unforeseen (often contradictory) consequences, none of which are determined by human intentionality. The trajectories of technological innovation cannot be adequately understood as being a product of an "inner logic" of development, but neither can they be adequately understood as the product of human intentions either. The situation is one of dialectical differentiation and development of trajectories that are not completely under human control, but also cannot be divorced from human decisions and interpretations. In his recent writings, Feenberg has acknowledged there are "natural and human constraints on technical development" and he has taken steps towards developing a dialectical theory of technology.[12] However, Feenberg admits to being unable to include science in this project, without imposing ideological and speculative ontological projects on science, and he assumes that the natural constraints on technical development are the remit of science.[13] As a result of this, his dialectical theory of technology is not developed significantly beyond a Marxist theory of dialectical materialism or Heilbroner's latest softening of technological determinism. However, as I argued in *On the Metaphysics of Experimental Physics*, the experimental natural sciences are technological activities that presuppose metaphysical precepts and are implicit in a grand societal experiment to construct the technological society. As I argued in *Modern Science and the Capriciousness of Nature*, this societal experiment presupposes the ideological and speculative ontological project of constructing an artificial world as an improvement of the human condition by promising to liberate humanity from the vulnerabilities and physical limitations imposed on us by the natural world. Once we take both these aspects into account, it becomes possible to understand "the scientific" as a subcategory of "the technical". This provides us with an historical interpretation of the possibility of science and technology upon which we can develop a substantive and critical theory of science and technology in relation to the question of whether the development of scientific research and technological development realises or distorts the ideals and values of the historical "civilizational project" upon which science and technology were premised as enlightening and emancipatory enterprises. This allows us to understand how "unintended consequences" and "technical limits" can emerge from the dialectics of technology, without adopting the Marxist ontological presumption that this dialectics is between social projects and the material world.[14]

Technical choices are emergent from a technological framework of structured choices, practices, opportunities, demands, challenges, and expectations, against a background of established means and ends that pre-exists and transcends particular social projects. Due to the underdeterminacy of complex systems, within which each change involves non-linear interactions, certainty is limited to special cases, within specific contextual bounds, and predictability is limited in scope, detail, and temporal extension. The understanding of the future behaviour of this kind of system is limited to technical judgements estimating probability, coherence, and stability, and involves explication of the assumptions, expectations, and selections, which underwrite the representations of the system required to make judgements of future behaviour. The understanding of future behaviour of such systems is interpretive and situated within historically developed frameworks of theory and practice, involving value-laden prior decisions about how these frameworks should be set up, developed, tested, and refined. Technical judgements regarding possible outcomes of any intervention within the system are themselves socially conditioned (paradigmatic) and contingent (subjective) judgements based on the projection of "closed system" representations over the "open system", utilising cultural meanings (albeit, often, subcultural, technically specialised meanings) in accordance with consensual understandings of intelligibility and plausibility in terms of a (sub)cultural stock of explanatory tropes and representations. Novel ideas are emergent from a background of paradigmatic conventions and assumptions about objective standards and methods, which are learned through technical education and usage. In this way, the bounds of technical rationality adapt through the dialectical differentiation and refinement of an historically structured continuum of developed theory and practice as and when they are applied to situated and particular problems. Even in "the closed system" of the laboratory, understood in the scientists own terms, scientific experience emerges from an historically structure continuum of developed and refined models, interpretations, standards, conventions, representations, methods, measurements, expectations, and levels of tolerance for error, imprecision, and simplification required to make any scientific observations.[15] This is further complicated by the fact that in the vast majority of cases of scientific investigation of open systems (outside the laboratory), scientific research is inconclusive and incomplete, therefore, scientists are unable to provide an "objective" assessment of probabilities and tendencies. Social factors, such as credibility and confidence come into play when scientists are required to provide estimates of likelihood, risks, and best possible outcomes, wherein the reputation of the scientists involved give weighting to the public acceptance of their estimates, and the evaluation of the rationality of any technical judgement is based on trust, even if cautiously or provisionally granted. On such an account "unintended consequences" and "technical limits" are the emergent products of structural incoherence and incompleteness of the technological framework within which technological activities and technical judgements are postulated, organised, refined, and developed, rather than corresponding to natural laws or mechanisms. Natural laws or mechanisms are cultural representations of the discovered alethic modalities of underdetermined technological activities involved in establishing coherent, stable, and communicable programmes of scientific research, experimentation, and technological innovation.

The decisions about what constitutes a rational technical choice emerge from an historical background of previous choices and trajectories, and, therefore, are socially contingent, but they are also limited by what is discovered to be the alethic modalities of technological activities. The possibilities, probabilities, necessities, certainties, and impossibilities of any technological activity are products of the structural interactions within the dialectically developed and differentiated technological framework, wherein all developments and differentiations are responses to the inconsistencies, incoherence, and incompleteness of the framework. These alethic modalities cannot be understood as objective laws, given that they involve human interventions and cannot be said to exist independently of particular modes of human engagement within the technological framework, which means that they cannot be divorced from human intentionality, but their outcomes are not controlled or determined by human intentionality either. While the actualities of our alethic modalities are shaped by historically conditioned and contingent choices, it is the capacity for the consequences of technical activity to frustrate and surprise us that shows that "the technical" is not circumscribed and limited by "the social" even through it is a subcategory of "the social". Machine performances can confound our expectations or lead to previously unimagined possibilities and can transform "the social" by bringing new "forces of production" and "their side-effects" into the world. This explains how the "the technical" can transform "the social" while also being contingent upon it in terms of social projects to construct a stable and coherent technological framework of interacting technological activities and objects. This has an inherently appropriative relation with Nature that transforms the natural beings under investigation or utilised as standing-reserve. The representations of the natural world produced by it (including the properties of natural phenomena and processes) are products of the technological activities and, therefore, technology mediates the scientific understanding of the natural world, including those aspects that we consider to be invariant and universal. The exploration, discovery, and representation of "the intrinsic properties" of technological systems are the products of technical activities that are dependent upon the contingent construction of the system, emergent from the technological framework, understood in technological terms from the onset, and, therefore, should be understood in terms of coherence relations rather than correspondence relations. Technological systems cannot be adequately understood only in terms of social construction of practices and discourse, nor can they be adequately understood in terms of a correspondence theory of objective properties. The system itself is an emergent property of its interactions within its environment, as well as the internal interactions of its components, that is contingent upon and conditions specific modes of social activity required to integrate and stabilise the system within its environment, but also has been developed and differentiated in response of prior "unintended consequences", such as unexpected resistances and inconsistencies. The success or failure of any effort to integrate and stabilise the system is itself a product of an experimental effort to converge the internal components into an ensemble of coherent relations to its environment, which includes other systems. From the onset, all such efforts are postulated to achieve specific goals and the understanding of the functions of the system's components cannot be abstracted from the expectations and aspirations emergent from the technological framework, without destroying their intelligibility. "The technical" is a subcategory of "the social", but the processes of determining the limits and uses of any technology producing representations of the cognitive-material consequences, or machine performances, which emerge in response to experimental interventions, developed to coherently understand these representations and activities in terms of a technological framework that precedes the experiment, as part of an ongoing grand societal experiment to construct the technological society as an more intelligible and predictable world than the natural world. It is implicit within the "civilizational project" of modernity, as Feenberg would put it, and can be understood and analysed in terms of critical theory providing that we include "unintended consequences" and "technical limits" as extra-social phenomena emergent from the ongoing construction of the technological framework. There is a dialectical dimension to the understanding and development of "the technical" in response to inconsistencies and incompleteness within the technological framework – an historical and sociological phenomenon – and, once established, scientific research and technological innovation take on their own agency, which effect (resists and empowers) future choices by bringing into the world new powers, which do not completely conform to human intentionality and expectations, and generate new inconsistencies and a further awareness of the incompleteness of the technological framework. In this respect, the technological framework can be understood as the ongoing and experimental concrete construction of the "civilizational project", while the implementation and further development of experiments and prototypes in the wider world have unforeseeable consequences for society and the natural world.

By acting as a mediator, the technological framework disciplines and organises human activity, policy decisions, strategy, and the choices about future scientific research and technological development, which are reiterated into refinements and modifications of the framework and the representations and technical codes it produces. This ongoing feedback process substantively transforms the contextual bounds and evolution of technical rationality and dialectically shapes the specific strategies and practices within the ongoing development and specialisation of technical systems and sciences, as well as their possible application to the societal organisation and practical problems of the "civilizational project". The reciprocal interplay between accommodations and resistance is inevitable when building technology into a structured environment. The overall development of any technology is the sum total of these particularities in context and its functions are bound to purposes, which imports both historical and social dimensions to its form, structure, and content. The full meaning of any technology cannot be grasped without providing a complete account of the particularities of its historical and social development and differentiation, but, given the interpretative relation with history and the unfolding character of the social, any such account will always be tentative and contestable. The understanding of technology is conditional upon how we understand the problems involved in interpreting history and society in general. This is further complicated when we also take into account that the unforeseen consequences of any technology (such as its "side effects"), which can only be known in hindsight because we live in an open-ended, changing, and complex world. Technical, historical, and sociological accounts of technology are necessary but insufficient for a complete understanding of technology because we do not have sufficient foresight to predict its future and we cannot derive its trajectory from universal natural or historical laws. We can only have a partial and dialectical understanding of any technology, within contextual bounds and temporal limits, providing us with limited knowledge and foresight based on a projection of our experience. This is not just an epistemological limit. The potential for any technology to be used for purposes unforeseen by its designers, in unexpected contexts and circumstances, which do not exist yet, shows that the consequences and meaning of science and technology are ontologically underdetermined in the sense that how we understand their usage and meaning is changed in relation to their development in particular contexts. The content, trajectories, and significance of any scientific research and technological innovation are an evolutionary trajectory, rather than fully determined by their initial design or internal logic. Innovations are the product of specialised techniques and devices imported from other technologies, which are transformed through non-linear interactions in order to create a stable and consistent technological framework by bringing together previously distinct devices, techniques, and technical expertise, and applying them in a novel context. Technical solutions to complex problems are always limited and incomplete because they constantly need to be modified, refined, and improved to deal with the complications and subtleties of these problems over and above the initial conception of those problems, as well as in response to any unforeseeable consequences which inevitably occur, but also because the problem is not isolated from other aspects in an open-ended, complex, and changing world. This shows that we need to recognise that our understanding of science and technology has both epistemological and ontological limits, but these limits are distinct from natural laws and mechanisms. Even if we possess a tentative and partial understanding of the historical and social development and differentiation of any aspect of scientific research or technological innovation, including a comprehensive account of all its known consequences and uses, we still cannot be certain that such an account is exhaustive, we cannot hope to predict all its possible consequences and uses, and we cannot hope to discover any "inner logic". This means that no one can claim technical expertise of how the whole framework operates, given that a total understanding of any technology is unavailable, and the specialised technical knowledge and experience required for technical expertise is divided throughout a community of specialists, without any overarching expertise in how that community should be organised. The meaning of scientific research and technological innovation are irresolvably ambiguous and the human condition is one wherein a veil of ignorance, pluralism, and the open-endedness of existence conspire to conceal the total meaning of any aspect of scientific research or technological innovation.

Modern experimental science is driven by the technological imperative to refine itself as the technological means to discover the truth about the natural world in terms of the instrumentality of natural phenomena. From its onset in the sixteenth century, it has been directed towards the discovery of the instrumentality of natural processes – releasing new powers and explaining how they work – and applying these discoveries to the pacification of Nature. The technological appropriation and transformation of natural phenomena into technological objects, machine performances, in order to prepare them for experimentation and practical application, while theoretically representing and understanding change in the natural world in terms of natural mechanisms and calculable systems has been socially connected through patronage, from the onset of the experimental sciences, with commercial, civic, and military ambitions. The subsequent historical trajectory of development, differentiation, and refinement of the scientific world-picture within which the production of prototypes is represented as the discovery of ontological depth, representing the historical stages of technological innovation as the stratified discovery of natural mechanisms, wherein subsequent technological innovations are used to explain preceding, has been epistemologically underwritten and legitimised by the practical successes that have arisen from the ongoing societal project of constructing the technological society to confront, dominate, exploit, pacify, and replace the natural world. As a realist pursuit, tested through ongoing technological activity, experimental science is simultaneously directed towards *techne* and *Ge-stell* to disclose Nature in abstract, non-sensuous terms, such as energy or entropy states, while also revealing its truth as intelligible causal explanations of its instrumentality. Within the scientific world-picture, objective truth is placed before us as something we can grasp, as something at our disposal. It is represented as the discovery of the most efficient mechanism at work during phenomenal changes. Given that scientific theories are "tested" in terms of their instrumentality within the technological framework of a scientific research programme, corroborated in equal measure to their usefulness for the further development of that programme and technological innovation in general, scientific truth is brought forth as tending towards *techne* (understood as the knowledge of the causal principles involved in making change happen) and as *standing-reserve* (in terms of their unspecified instrumentality for future use). Technological innovation occurs within this culturally situated and historically developed technological framework of the stratification and refinement of machine families and kinds.[16] A substantive and critical theory can analyse scientific research and technological development as dialectical rather than continuous processes, which means that even though their trajectories are stratified, requiring the development of basic technologies before developing more complicated ones, the further development of these trajectories occurs through non-linear relations within an environment and, therefore, given that changes in that environment will change the non-linear relations of further technological development and the outcomes of their interactions remain irresolvably ambiguous and unpredictable. For example, the science of mechanics was developed from the ancient and medieval projects of mathematically describing the mechanical motion of the six simple machines, which constituted a machine kind from which the entire machine family of mechanical devices could be constructed. The nineteenth century development of electromagnetic machines involved the innovation of a new machine kind which cannot be reduced to the six simple machines, requiring a different set of basic motions and their representational techniques. Machine kinds are innovated by bringing together and integrating heterogeneous machine kinds from disparate machine families, as an experimental creative process of converging and stabilising machine performances, and, therefore, the emergence of new machine kinds requires its predecessors, but it cannot be explained (reductively) in terms of them even in hindsight and therefore could not be predicted before their serendipitous discovery. The representations of the basic motions of any machine kind are the product of complex social processes to stabilise, reproduce, and communicate them.[17] Prior judgements about how to proceed with scientific research and technological development constrain the direction of their further development and complex technologies require simpler technologies in order to be possible, but the assertion that the implication of this state of affairs is that technological innovation occurs in accordance with a continuous and predictable "inner logic" is one that can only be made in hindsight by leaving the contingencies involved out of the account. For example, while the nuclear fusion reactor clearly requires an ongoing stratified technological development to create the conditions for its existence, such as the development of mining, chemistry, electricity generation, and electromagnetic focussing technologies, alongside the theoretical development of High Energy Physics and hydrogen containment technologies, the process by which the nuclear reactor could be developed cannot be logically derived from any of these strata or theories because the context of problems and demands, within which technological convergence occurs, cannot be logically derived from the history of scientific research and technological development. Rather than being a continuous and predictable product of "inner logic", the situation is more of an evolving system, wherein strata are states of punctuated equilibrium, where long periods of stability are followed by brief periods of radical change that lead to new problems, contradictions, incoherence, and instability, thereafter a new order of stability is established by dialectically solving these problems, resolving the contradictions, achieving coherence, and producing stability, but generating new problems. Each innovation has emerged as a response to a cultural framework of values, intentions, expectations, problems, and standards, but the technical processes involved in making that response are experimental, non-linear, extra-social, and unpredictable. When new strata of machine kinds emerge from within an accumulated stock of technical knowledge and resources, they create new alethic modalities, problems, and challenges for future development of the technical framework, often displaying a high degree of ambiguity, and, at all stages, human choice and judgement is involved in their implementation and development. These lead to unforeseeable demands and constraints which shift the boundaries of technical rationality. This is not simply an epistemological limit, but is a consequence of the ontological limit in a complex, changing, and open-ended world. This is an inherently creative process wherein the experimenters discover how to stabilise, reproduce, and communicate the technical process. This means that each stratum in the stratified history of research and development involves the discovery of new rules, but the next stratum cannot be known in advance because the conditions for its creation are the dialectical outcomes of efforts to complete and stabilise the technological framework in accordance with these new rules in a world that does not conform to our intentions and expectations. While every technology has a structured history, its exact trajectory of development requires the creative synthesis of heterogeneous elements and, therefore, it can only be determined in hindsight. Our explanation of how this trajectory is possible – in terms of the exercise of natural mechanisms in accordance with natural laws – is itself a product of communicative and technical activity. Of course, we can imagine a futuristic machine, such as a teleportation device, represented in terms of quantum theory, but its invention will involve the discovery of how to make its components and successfully integrate them into a stable machine performance, within a wider world, which we at present have no idea how to do or even if it can be done. We cannot predict the sequence of inventions and discoveries that could lead to the invention of a prototype, let alone predict its consequences and evolution subsequent to the implementation and development of that device in the world, and, should such a device and its supporting infrastructure be invented, the continuity and predictability of the trajectory of its technological development would be a product of historical hindsight and would not corroborate the quantum theory used in its original conception. The development of this machine would remain incomplete and comprised of underdetermined and heterogeneous components (each with their own historical development) which interact with each other and their environment in unpredictable and discontinuous ways when efforts are made by disparate (sometimes competing) social groups to converge science and technology to solve different problems in a complex, open-ended, and changing world.

By examining how technological development places constraints and demands on human choices, without appealing to some kind of technological determinism, in terms of the dialectical development of the technological framework constructed out of heterogeneous elements, we can see how technological development tends towards some directions while resisting others.[18] It is in reference to the technological framework that we can see how social, political, and economic criteria shape the bounds of technical rationality, which shapes technological development, and, in turn, by bringing new possibilities and problems into the world, shapes the social, political, and economic landscape of expectations, possibilities, ambitions, and limitations. Following Heidegger, Ellul, and Habermas, Feenberg was critical of the abstraction of the technical from its total set of relations, objects, and systems, all of which comprise its being, because it decontextualises and reduces the technical, misrepresenting the concrete in all its particulars, but he was critical of Heidegger, Ellul, and Habermas for considering the essence of technology only in terms of primary instrumentation. Feenberg argued that primary instrumentation provides the basic technical relations, as a framework, which only yields an actual technique or device once these technical relations are integrated within the natural, social, and technical environments that support its functioning. However, Feenberg tends to oversimplify this process, as if a framework of technical relations can be constructed independently of the processes required to integrate them into the world. It is only through the processes developed through attempts to integrate any technique or device into the world, as a stable and consistent technique or device, that it can achieve technical relations and functional constitution at all. Technical relations and functional constitution are actually emergent from the process of constructing the technological framework within the world, which modifies the intentions and expectations that gives functionality its meaning through applicability in context. These considerations, when applied to Feenberg's characterisation of technology in terms of its *functional constitution* (its primary instrumentation) and its *realisation* in actual technical practices and systems (its secondary instrumentation), show that there remains irresolvable ambiguity between these two aspects as each dialectically transforms the other during the implementation and development of any technology in the world. The functional constitution is comprised of realisations, while realisations obtain their meaning in terms of their functional constitution; both aspects continually inform and constrain each other as the whole technological framework is developed and differentiated. Hence, the functional constitution of primary instrumentation cannot be identified and analysed, as if it were separate from the secondary instrumentation of its realisation, during the dialectical process of technological development, without distorting our understanding of that process. Consequently, the ambiguity between the primary and secondary aspects of instrumentation highlights the experimental character of every technology, highlighting how the development and differentiation of the technological framework is ongoing and perpetually incomplete, requiring continuous modifications and refinements during the processes of attempting to integrate it into the world, which is itself changed, however slightly, by those attempts, as an ongoing affirmation of the practical value of science and technology for the "civilizational project" of constructing a technological society as an improvement upon the natural world. The trajectories of technological development are the consequences of an historical struggle to establish and disseminate human choices, values, goals, and meanings by resisting and adapting to natural forces and events beyond our control, but the outcomes of these struggles are encoded into the form, structure, and content of "the technical" and, thereby, constrain and direct future choices, values, goals, and meanings. The horizon of possibilities and the means to reach it are shaped to satisfy political, economic, and social aspirations, which technology promises to achieve, but, in turn, new aspirations are only possible because science and technology bring new possibilities into the world. The realisation of these new possibilities dialectically shapes the potential trajectories of further technological development. Technical judgements are bounded human choices, with aesthetic and ethical aspects, but these choices are constrained, resisted, and directed within the technological framework of research and development, which imposes demands, challenges, and expectations upon human beings once we consent to work within that framework. Technical judgements are disciplined human choices made within the technological framework in response to the demands of that framework. These demands place limits on the possibilities for technical codes.

Choosing between civilisations

The advocates of instrumentalism treat technology as a well understood and a subservient value-neutral means to rationally satisfy ends or goals established in other social spheres, e.g. politics or culture; whereas, the advocates of substantive theories describes technology as an autonomous phenomenon, overriding all traditional or non-technical values, shaping both humanity and the natural world, and disseminating and transforming ends and means, as both an environment and way of life. On the instrumentalists' account, the design and structure of technology is not an issue in political debate, which, instead, needs only to focus on its consequences, suitability, and implementation. According to Feenberg, advocates of substantive theories of technology have rightly criticised the instrumentalists, but, on the substantivists' account, we are condemned either to pursue its advance toward dystopia or to return to a more traditional or simple (pre-modern) way of life. According to Feenberg, for both theories, the battleground is a moral and political debate about the destiny of humanity and the choice between civilisations, with the instrumentalists being for modernity and the substantivists being against it. In making his criticisms of the substantive theories of Heidegger, Ellul, and Marcuse, Feenberg argued that modern technology is not simply a goal-orientated pursuit of efficiency, which he termed as "primary instrumentation", but constitutes an essential dimension of the struggle for a humane world, conducive to human life, which has been distorted by industrial capitalism. As Feenberg acknowledged, Marx proposed the idea that the control of the means of production by workers would allow the redesign of technology to apply high levels of skill to production, which, in itself, would lead to deep changes in education, politics, and social life. Capitalism imposes control from above, structuring labour and technology in order to permit the perpetual replacement of skilled workers with cheaper and more controllable unskilled workers. Thus the economic interests of the owners of the means of production determine the design and structure of technology. If this is true then the prevalence of the unskilled and uneducated workers is not a consequence of technological innovation *per se*, but is rather a product of social structures that suppress and control the education and democratic potential of the labour force in order to increase the power and wealth of the social elite. It is because the labour force is reduced to expendable components, on a principle of reducing the exchange-value of labour, whilst the public education system is geared to reproducing such a labour force, that the democratic potential and class-consciousness of the citizens of industrial society are suppressed. On such a view, the democratisation of the technological society would be contingent upon a radical and egalitarian redistribution of knowledge, wealth, and social power within society. The problem of the democratisation of science and technology is not a problem with science and technology *per se*, but is a problem with social organisation and the way that the technological structures of industrial production enforce and reproduce the social structures of industrial society. The centrality of the concept of "efficiency" (the maximum amount of work in the minimum of time with the minimum amount of resources) to the social organisation of science and technology prevents democratic participation and gears industrial processes to the satisfaction of short-term economic and political goals (i.e. the reduction of production costs and the maximisation of production rate), at the expense of long-term economic and political goals, such as social development and environmental sustainability. As Feenberg argues, we must question the presumption that a democratically controlled economy would lead to an inefficient and incoherent organisation of labour, technology, and resources. We need to question the presumed trade-off between "virtue and prosperity", as Feenberg puts it, by raising the question of whose prosperity? Feenberg asserts that the exclusion of the vast majority from participation in this decision is the underlying cause of many of our problems, and a good society should enlarge the personal freedom of its members while enabling them to participate effectively in a widening range of public activities. He argues that a profound democratic transformation of modern industrial society will resolve these problems.

The inherent instability of the networks that comprise any technology allows new possibilities to emerge, as well as conflicts and resistances, and challenges the current social organisation of technology and its strategies of development and implementation. The technological society involves a complex of inter-dependent networks, systems, and organisations. Changing one technological system will involve changing others and re-organising much of society. The democratisation of the technological society is an enormous task, requiring a radical deconstruction and transformation of many things that we take for granted. Feenberg rejected claims that the "lay citizenry" are insufficiently educated to participate in scientific research and technological innovation. As Feenberg argued, producing a surplus of education and radically redistributing technical knowledge throughout society is necessary for the democratisation of society.[19] In my view, Feenberg is quite right about this, but we also need to include scientific knowledge, as well as broaden the understanding and participation of both scientists and the "lay citizenry" in the debate about the meaning and purpose of science and technology. It is crucial for the democratisation of the technological society that we all learn much more about how the technologies we take for granted work, their limits, their effects, their potentials, and how they could be developed for our benefit. It is also important that we all learn more about basic technologies, such as plumbing, woodcrafts, electrics, horticulture, mechanics, etc., in order to increase our own understanding of the basic infrastructure of our lifeworld, and also to increase our own self-reliance and technoscientific imagination about how we can develop our homes and community for our benefit. It is also essential that we all try to understand radically novel technological innovations and how we can use and adapt them for the benefit of our communities and lives. We need to learn how to actively participate in the transformation and re-invention of many technologies in order to radically democratise their total definition by transforming the way that they are used and integrated into society. The democratic revolution is one that involves people cooperating to radically transform the "civilizational project" into something that empowers and benefits people and communities. Every technology has ambiguities and contradictions in its total definition and it is important for people to identify and analyse these in order to transform them into areas of flexibility that can be exploited in pluralistic ways. When integrated technological systems are mutually transformative and empowering then they become interdependent. A failure in one system can radically transform and disempower the whole network of interdependent systems. This emphasises the need for maintaining in-built local independence and self-sufficiency through the whole democratic, bottom-up process of constructing large, interconnected systems. Thus should the system fail, at any level, then local communities can fall back on readily available alternatives and immediately cooperate with other communities to begin the process of attempting to discover and correct the failure. All technological systems have to be democratically integrated as part of a pluralistic social fabric of ordinary, everyday life. The citizens of a genuine democracy must assert and exercise their right to know, to question, and participate in the processes of the selection, development, innovation, and implementation of new technologies within society. I agree with Feenberg that the interpretive flexibility of any technology provides a pluralistic basis upon which the democratisation of technology can proceed. Indeed, democratic participation in the development and implementation of technology can awaken an awareness of the unrealised potential of those technologies. In Marcuse's terms, this provides the basis for a dialectical critique of the actuality of the current strategies for technological development and implementation in relation to the historical development of our ideals and potential. However, unless we are aware of the metaphysical foundation of the technological society and the moral imperative of the societal gamble, we maintain an uncritical complicity in the technocratic ideology, even if we reject it as totalitarian. Without this awareness, we will be perpetually condemned to reproduce the social conformity to the representation of technical rationality as being the rational basis for choosing between strategies of technological development, even if we reject the claim that technical rationality is the best means to understand how to achieve human well-being and the good life. If we are to identify and realise alternative forms of rationality *over and above* technical rationality then we must critically examine the metaphysical presuppositions upon which the rationality of technical rationality depends. Feenberg correctly argued that it is essential for the possibility of genuine democracy that technical strategies serve communicative strategies. However, unless we examine the fundamental presuppositions upon which the development of our society depends, in terms of the presupposed goals and accounts of human well-being and the good life, then our communicative strategies will remain irrational.

Unless the metaphysical foundation of the social acceptance of the authority of technoscientists to write and interpret the technical codes of society is successfully challenged, then the societal gamble remains unquestioned. Therefore, on the question of who is best to make technical decisions, the rational choice will always remain on the side of technical experts, whereas, once the rationality of the societal gamble is brought into question, then the goodness of the bounded technical rational solution is also brought into question. Once this happens then the public is at liberty to rationally question the rationality of any technical decision in any given context, and also to undermine the previously absolute authority of the technical experts to make decisions regarding the development of technology because it is no longer self-evidently clear that technical rationality will provide the best solution. It becomes only one solution among others about what is best for human well-being and creates a discursive space for non-technical forms of rationality to be involved in the decision-making process about community development. But, while rationality is defined in terms of bounded technical rationality (even if it is rejected) then there is no rational basis for non-technical participation in technical decision-making. Without this rational basis for a critical re-evaluation of the rationality of technical rationality, any "democratisation" of the technological society, such as Feenberg's so-called "deep democratisation" is itself irrational. Feenberg has failed to address the underlying reasons why technical experts are represented as serving human interests by providing the means to satisfy those interests, and, hence, he has not addressed the underlying metaphysics of the technocratic ideology. Thus his call for democratic rationalisation does not tackle how technical rationality has historical pre-eminence in modern society. As a consequence, while we may well sympathise with his assumption that democracy is an intrinsic good, Feenberg's call for the democratisation of the technological society is without rational foundation. It is essential that we discuss the practical value of pluralistic and diverse democratic participation for the purpose of understanding the potential for the rational and sustainable development of society.

Feenberg argued that to struggle for different technologies is to struggle for a different society.[20] If we wish to construct a genuinely stable and sustainable society, we need to question the technological structures, productive relations, patterns of consumption, and device paradigms embodied in our everyday experiences and practices. This involves challenging the presuppositions of the *status quo* by offering genuine alternatives which can provide a practical means to sustain and develop human existence in accordance to an alternative vision of civilisation. It involves choosing between "civilizational projects". Feenberg argued that technocracy is an ideology that suppresses the call for democratic participation in this choice. But, how did he explain the ability of technocratic ideology to do this? He argued that devices enforce moral obligations in our society and embody the values that human beings have built into the functionality of these devices and their context of usage.[21] In my view, his answer is unsatisfactory because, even if it is correct, it does not explain why this process of embodiment is directed in accordance with the technocratic ideology, nor does it explain how technocracy is able to impose its decision regarding the use of technical codes to expand its system of hierarchical control. Feenberg has only pointed out that technocratic decisions are embodied in technical codes, but we need to explain how that embodiment was possible in the first place. We need to explain the historical conditions and contingencies that transformed society in a way that empowered the technocratic hierarchy, as a legitimate authority, in the first instance. It is not enough to show how the technocratic ideology assumes technological determinism, as Feenberg does, but we also need to explain how this assumption was culturally meaningful, rhetorically intelligible, and socially acceptable. As I argued in *On the Metaphysics of Experimental Physics*, experimental sciences, such as physics, are premised on a set of metaphysical precepts which secured logical connections between epistemology, refinements of scientific world-picture, and the practical successes of technological innovation. This was ontologically connected to the societal gamble in the goodness of the societal construction of the technological society as an improvement upon the natural world. Once the faith in the societal gamble had been culturally established as being rational (or self-evident) then the legitimacy of the whole project of the construction of the technological society, including the experimental sciences, could be rhetorically justified upon technological determinism. After technological determinism had become implicitly accepted, through the early successes of the mathematical science of mechanics and the new sciences, then the mechanistic world-view became accepted as the sole basis for knowledge of the natural world and the rational development of society involved using the technical expertise of how to apply objective scientific knowledge to satisfying human needs. As a series of research programmes, science is further developed and differentiated as an ongoing series of specialisations dealing with specific kinds of machines and techniques, further developing and differentiating them within specific strata of the technological framework. In this way, those that funded scientific research and technological development through patronage were able to translate their hierarchical social power into a technocratic ideology, wherein ownership of the means for institutionalising the specialisation of scientific research and technological innovation equated to control over the social structures of the deliberation and decision-making process about how best to implement and develop science and technology. It is not only necessary that we challenge the assumptions that underwrite technological determinism, but we subject the whole project of constructing the technological society – as a "civilizational project" – to public scrutiny. Not only should we question the technical rationality of technocratic decisions, by asking whose interests they serve, but we also need to question the broader rationality of technical rationality by examining the historical and sociological development of this "civilizational project" and question the relation between those decisions the implicit vision of the ideal society that technical rationality presupposes. This allows us to recover the potential for constructing a rational, egalitarian, and libertarian society, should we so wish to construct such a society, by critically comparing the actualities of societal development in relation to this recovered potentiality. This involves questioning the rationality of the societal gamble by critically examining the goodness of the technological society as an ideal society, as well as the normative representations of the human good life and well-being that such an ideal entails, alongside questioning whether the current trajectories tend towards this ideal or distort it. Of course, questioning the rationality of the societal gamble will involve the philosophical development of accounts of how technology relates to human potentiality, involving an account of the good life and human well-being, in order for empirical statements about the benefits and harms of particular technologies to be meaningful. Given that there is an absence of any universally accepted account of the good life, we need to recognise that there is a plurality of contenders for such an account and take this plurality and diversity into account in developing an understanding of how to proceed rationally. In this respect, the development of a non-ontological account of human potential, upon which a critical theory of science and technology can be developed, must be achieved as part of a critical and public deliberation of the actuality of societal development in comparison to historically recovered ideals, aspirations, and possibilities, while also incorporating "the veil of ignorance" imposed upon us by epistemological and ontological limits. Given the practical need for a decision about how best to proceed, the plurality of contenders is a testament to human ingenuity and experimentalism, but the decision between them will have a substantive effect on the quality of human life, given that some activities and conditions are good or bad for us independently of our knowledge and description of them, and, therefore, every proposal is worthy of consideration, but some proposals will lead to better consequences than others. This is a condition for rational decisions based on reasoned deliberation to be related to estimations of the practical value of any proposal. By embodying the plurality of contenders in the deliberative process of deciding societal development, we can maximise the chances of discovering how to live well through trial and error and also learning from the experiences of others. As I argued in *Modern Science and the Capriciousness of Nature*, once we recognise that every account of the human good life is premised upon an experimental way of life, embodied in the discourse and practices of everyday activities, the practical value of such an account must be discovered through *a posteriori* evaluations of whether human beings flourish or flounder by living thus within particular communities. The details of any such evaluation must be a matter for local democratic deliberation and agreement by the same people who are likely to experience the consequences of their decisions, but regardless of any consensus, such decisions should remain pragmatic and tentative, given that they might be mistaken or future events may well lead to a reversal of fortunes and the need to re-evaluate our interpretations of any way of life. Hence, societal pluralism and diversity have practical value for society as a whole, given that, by allowing different communities to democratically explore different ways of life and learn from each others' experiences, it limits the societal damage which would result from local errors in judgement regarding the good life and how to live it, while it increases the chances that the good life could be discovered by chance. It provides a buffer against the "unintended consequences" of technological development, and prepares us to rapidly accommodate unforeseen events, such as social changes or natural disasters, by increasing the adaptability of society as a whole through the decentralisation of the technological development of society through the democratisation of society at the level of local communities.

The societal gamble involves a faith that science will lead to controllable and predictable existence, but it does not require such an existence to begin its attempts to pacify existence. Science is a cultural response to an uncontrollable and unpredictable world – a capricious natural world – and democracy is a societal response to the absence of universally agreed epistemological and moral standards, including those for science. Even though directed towards the enhancement of predictability and control – the pacification of existence and the satisfaction of the societal gamble – the implementation and further development of scientific research and technological innovation involves the realisation and exercise of technological power in the world that changes the world in unpredictable and uncontrollable ways. This actually increases the level of complexity and necessitates the emergence of the "vigilance" of democratic participation in equal measure to the awareness that there is an absence of any certainty about how to proceed to solve any given problems, nor is there any universal agreement about what the problems are, how they should be prioritised, and what their satisfactory solution should be. Democratic participation is the political expression of a pluralistic and epistemological response to the ontological ambiguity of human action and the consciousness of human fallibility in an open-ended, changing and complex world. As a result, democratic participation remains the best way of recovering the practical value of deliberative reason by maximising the level of decentralised societal plurality, dissent, and criticism. Any technocratic distortion or suppression of dissent and creativity will inevitably distort or suppress the societal capacity for rational societal development. Due to its structured identification of all societal aspects of development as being a series of technical problems and the methodological reduction of their possible solution to the establishment of a set of techniques or paradigmatic horizon of research and development, anything outside this framework is represented as either subjective or irrational, and criteria for evaluating the practical value of proposals are imposed *a priori*. This inductively assumes that the future will resemble the past and reduces the societal capacity to accommodate and adapt to unforeseeable events. The necessity of the subordination of scientific research and technological innovation to democratic deliberation and decision-making is a direct result of the increase in ambiguity that has emerged from the scientific effort to achieve closure, control, and certainty in an open-ended, complex and changing world. This need for democratic participation is accentuated by the awareness of the inability to reduce human values and goods to objects of scientific research and technological manipulation. Hence, communicative action cannot be reduced to strategic instrumentalism. Of course the public perception of science as a privileged means of providing closure and definitive answers in the face of ever increasing complexity and ambiguity has lead to the perpetual deferment of deliberative and decision-making processes. The public perpetually awaits the outcome of "scientific consensus", as conveyed through media, while the debate about public goods becomes reduced to the debate between scientists about the scientific validity of the available information. For example, media debates about the validity of scientific evidence for global warming have become a surrogate for deeper debates about the preservation of the natural world and the consequences of industrialisation, as well as its environmental impact and the need for its regulation. The opportunities for important societal debates about respect for the natural world, public goods, social justice, and human well-being have been reduced into a series of propaganda exchanges about the facts and credibility of opposing groups. Thus it is quite unreasonable to promote the democratisation of science without promoting the democratisation of wider society.

The benefits of pluralism can only be achieved if there is an effort to communicate evaluations of the good life across society, allowing communities and individuals to learn from others, as well as help others learn how to live life well via processes of trial and error. Should pluralism degenerate into relativism then all its benefits would be lost. In this sense, even though we do not agree about what constitutes the best course of action to achieve the good life, or human well-being, the meaning of communication for democratic participation implies that we accept that there is a possibility of discovering benefits and harms, making errors of judgement, learning from the mistakes and successes of others', and that the consequences of our actions are not determined by our intentions. Tension arises when science is aimed at delivering benefits to society through the achievement of predictive certainty and technological control, while the world and democracy involve uncontrollable and unpredictable relations and processes. Does this tension reveal an inherent incompatibility between science and democracy? Any incompatibility between science and democracy is not due to the nature of science, technology, nor democracy, but is due to how science is organised within wider society; it is an incompatibility between democracy and a hierarchical society within which science is directed to satisfying the interests of economic, commercial, and military ambitions of a social elite, without regard for the public good. The problem of the democratisation of the technological society is a consequence of the antidemocratic structures of society. Science has not been organised in such a way as to incorporate democratic principles and ideals, such as equality of access and participation, nor has its research priorities been directed towards benefiting humanity in general. Instead of adhering to the Enlightenment idea of developing universal scientific knowledge for the universal benefit of humanity, the Western development of science has reduced knowledge to a commodity produced for the purpose of generating advantage of one group of human beings over another. Scientific research is increasingly directed and constrained in accordance with the interests of national defence and private industry – directed to secure increasingly marginal advantages over competitors (real and imagined) by increasing the costs of competing. Natural science has indeed become a techno-science within the infamous industrial-university-military complex. Large-scale scientific projects, such as the construction and operation of particle accelerators for High Energy Physics, may well be internally directed towards "pure research", but such projects have been funded by governments to provide valuable testing facilities for technological innovations, some of which have immediate military or commercial applications, and also to train scientists, some of whom become directly involved in weapons research or industry. Indeed, physics has dual aspects as *Ge-stell* and *techne*, bound up with ongoing technological innovation, to satisfy the demands and challenges of wider society, while its practitioners remain concerned with discovering the causal laws that explain how these technological innovations are possible. Hence, scientists can act as if they are autonomous and directed towards "pure research" even when their work is structured within a technological framework directed towards and bounded by political and economic demands. The commercialisation of science by multinational corporations in the development of biotechnology, for example, has intensified the anti-democratic tendency towards resisting and suppressing the societal opportunities to use science to restrict the introduction of its products into the wider world by concerned citizens and professional politicians. The high degree of commercialisation, performed at an accelerated rate in order to secure market advantage and patents, introduces a high degree of competition and secrecy among increasingly specialised scientists, which reduces the level of participation in public deliberation by other scientists, as well as reducing the availability of their scientific work to public scrutiny. Technological rationality, directed in accordance with the capitalist economic imperative, conceals the framework of social interests which organise scientific research and technological innovation in society, and suppresses the openness of scientific inquiry as well. Once we take this into account, we are able to recognise that there is not any essential incompatibility between science, technology, and democracy, but there is an incompatibility between democracy, science, and the contemporary state of the organisation of society itself. If this is true, the democratisation of science and technology would follow in due course from the democratisation of society.

The history of the industrialisation and urbanisation of society has resulted in increased levels of consumption, as well as a greatly increased range of products available to consumers, but it has resulted in the increased alienation of the vast majority from the administration and legislation of society. While freeing human beings from the material limitations of human existence, technology has also enslaved human beings within technical and economic systems of production, consumption, and exchange. All aspects of human life, including work, leisure, arts, culture, media, science, politics, sexuality, and even religion have become increasingly mediated by these technical and economic systems. Opportunities for spontaneous decisions about desires and action are suppressed by the system of ends and means, which juxtaposes clearly specified results with appropriate procedures. The systematisation of human existence objectifies it into an ensemble of techniques and products. Within the technological society, human relations are objectified as relations of production and consumption in two ways. The capitalist economic imperative objectifies human relations by quantifying them in terms of exchanges of goods and commodities; the technological imperative objectifies human relations by standardising them in terms of reproducible techniques and their results. Under either imperative, human relations become mediated by and incorporated into the structures of systems of production and consumption. These systems are the media through which the inequalities of power and wealth are reproduced and intensified. Feenberg proposes that critical theory of technology is the basis for a rational technological politics for choosing between civilisations. A critical theory of technology promises to bridge the gap between resignation and utopia, by providing a means of analysis and communication between the radical intelligentsia and technical experts. It allows us to explain how modern technology can be redesigned to adapt it to the needs of a free and rational society. He argued that, with the exception of Lukács and Marcuse, Marxist critics of technology failed to explain the new relation to Nature and provide a concrete conception of the "new technology" implied and required by their critiques. He applauded the efforts of the Frankfurt School and Lukács for showing that a critical theory of technology was possible, as an ongoing critical reflection and analysis of social domination, which demands a radical democratic advance and liberation of the technological base of modern society. Feenberg criticised Ellul, Heidegger, and Habermas for proposing an overly abstract theory of technology, rather than paying attention to the specific content of technical decision in context. However, in my view, Feenberg has missed their crucial critical point. Ellul, Heidegger, and Habermas attempted to raise consciousness of the extent that modern society is based on abstracts, devoid of specific content, which are disseminated as if they were real objects. It is the extent that human interactions and reflections have been replaced or mediated by abstractions that is of paramount concern. By ignoring this aspect of modern society, in favour of focussing specific choices between technologies, as if the technological society really allows us such a choice, Feenberg has presupposed the reality of an abstract notion of technology as being something that is at our disposal, as if its direction and content were simply a matter of decision. This represented the extent that we are compelled to participate in the development of technological society as being a matter of an abstract "democratic choice" between civilisations expressed in a democratic choice between technical codes. This fails to address that this "democratic choice", in such a context, is an illusion. It is the absence of genuine opportunities for democratic choice that is the problem. It is more the case that our concern should be a democratisation of society, rather than a democratisation of science and technology, as if we already lived in a democratic society and simply forgot to apply it to technological development. Feenberg asserted that "we have delegated the power to decide where and how we live, what kinds of food we eat, how we communicate, are entertained, healed, and so on".[22] But this assumption is questionable because it seems that most of us (if not all of us) never had this power in the first instance – let alone delegate it. The majority of people do not have any social power to choose between alternative technical codes and hence do not have a choice in which of any possible civilisations to construct. It is almost as if Feenberg supposed that we simply are unaware of the fact that if we choose different technologies then we would construct a different civilisation. Are we to believe that this choice has been available to us from the onset and we simply were unaware of it? In a Sartrean existential sense, perhaps Feenberg is right, but acts of identifying, analysing, and deconstructing the technical codes built into our economic and political systems, which are designed to resist or direct public participation in the construction of the technological society, are not sufficient unless we confront the antidemocratic and authoritarian hegemony of the economic elite. The problem for the democratisation of society is that the vast majority of people do not have any opportunities for participation in deciding the "civilizational project", but are treated as functionaries and labourers by the technologically empowered economic elite. Until this fundamental inequality within our society is addressed, the technical codes that are built into bureaucracies and the political process will remain unchanged. Indeed, within the technological society, the technological imperative is a fundamental impetus of social change, but unless we identify the obstacles to democratic participation in social change then this awareness achieves nothing. Without an ongoing struggle to democratise society, there is no possibility of rewriting the technical codes in such a way as to empower democratic participation. Feenberg considered the question of the democratisation of technology as if it were an epistemological problem of knowing our democratic possibilities – as if we had just been ignorant and lazy about the exercise of a power we had all the time – whereas the reality of our social being may well be that we do not have any democratic possibilities at all, and, therefore the democratisation of society is the ontological, political problem of the radical democratisation of our social being, and may well be a much deeper problem than Feenberg suggests, whereas Feenberg treats it as simply a matter of participation in the construction of the technological society. If this is true then Feenberg's argument is simply designed to raise our consciousness, in order to awaken our latent power of choice, whereas, it may well be the case that a more radical ontological transformation of social being is needed in order to recover the possibility of a democratic revolution in the collective participation in the alternative construction of our own communities, without waiting for permission and instruction from politicians and technocrats. Feenberg was inspired by Benjamin Barber's model of strong democracy, but due to his lack of any critical account of the rationality of the technological society (beyond merely pointing out that it is socially contingent), his own model of democracy remains limited to institutionalised reform of public participation in the technological development of society, modifying technical codes and electoral controls on technical institutions.[23] Admittedly this does bring wider criteria to the process and is important, but it lacks any developed model of the framework through which the public can empower itself and transform how such decisions should be made, except through a political process of reforming the current system.

However, once we place the origin of modern technology into its historical context, as a societal gamble in the "civilizational project" of constructing the technological society, then we can see how Feenberg underestimated the extent that the struggle for a humane world, conducive to human life, has become transformed into the goal-orientated pursuit of maximising efficiency, understood either in terms of the maximisation of productivity or the minimisation of costs. This is not simply the outcome of an epistemic fallacy, but is the outcome of a fundamental orientation towards the human condition. As I argued in *Modern Science and the Capriciousness of Nature*, the transformation of liberal capitalism to industrial capitalism, during the nineteenth and early twentieth centuries, subordinated the concept of technical efficiency in accordance with economic efficiency. This resulted in a series of phase shifts between the technological imperative and the capitalist economic imperative where the former would dominate to increase productivity and the latter would subsequently dominate once a profitable level of productivity had been achieved (reducing labour costs and suppressing further refinements of technical efficiency until a sufficient return of investments had been achieved). Cycles of capitalisation, overproduction and redundancy, termed by Marx as crisis-cycles, can be understood in terms of the phase shifts between these two imperatives. The technological imperative dominates during the development of high productivity, while the capitalist economic imperative dominates when the productive capacities of industry and innovations threaten profitability (leading to inbuilt obsolescence and the suppression of research). The demands of capitalisation and its crisis-cycles have subordinated the technological imperative to the demands of mass production and consumption. Feenberg argued that industrial capitalism develops "primary instrumentalization" strips technology bare of values and social context. However, Feenberg has only analysed capitalism in terms of relations of production, but if we also analyse it in terms of consumerism then we can see how industrial capitalism transformed "secondary instrumentalization" into the "device paradigm". Although Feenberg is correct to analyse the implementation and development of technical codes as historical and sociological phenomena, he has neglected to attend to the way that "secondary instrumentalization" is used to disseminate and secure the cultural acceptance of "primary instrumentalization" by conflating the societal gamble and the device paradigm. The technological imperative and bounded technical rationality promote specific values and social relations, which disseminate, reproduce, and universalise these values and relations as cultures incorporate imported devices and technologies. This has a much deeper political purpose and cultural meaning than Feenberg recognises. It is the reification of specific modes of social being as being the realisation of "universal rational man", wherein technological determinism justifies economic globalisation. Technological determinism is deeply implicated in political and ontological representations of the "civilizational project" of universalising specific modes and structures of social being in order to construct a global economy and neo-liberalism. It is not simply the case that advocates of technological determinism are ignorant of the dependence of technology on social relations, but it is much more important to address the way that determinists represent the choices of technological development as necessarily being beneficial for humanity in totality, as necessarily leading us to a better society, the best of all possible worlds, free from the capriciousness of Nature.

Feenberg called for greater public involvement in the design process; opening the design process to wider criteria than efficiency, risk, and cost. Public involvement should not be restricted to imposing ethical concerns regarding the obligations and limits of scientific research and technological innovation because democratising substantive decisions about how to implement science and technology as public goods should be a public concern too. This is an important point. Decisions between different kinds of technologies involve moral decisions regarding our obligations and limitations, as well as technical ones regarding efficiency.[24] Feenberg terms public participation in the development of new technologies from their earliest stage as "reflexive design".[25] It may well be the case that in choosing between traditional practices and modern technologies we may well find ourselves choosing between the wealth of engagement and the affluence of the device paradigm – this may well also be the case when choosing between modern technologies that are cost-effective in the short term but ultimately unsustainable and polluting, such as fossil fuel burning electricity generation technologies, and sustainable technologies that are expensive in the short term, but clean and reusable, such as solar power cells or wind turbines. These choices between short term affluence and long term wealth are fundamental decisions about the directions of development of society. Such choices involve decisions about what the technology is for, but they also involve decisions about what society is for. We need to focus on technology as a focal thing – developing technologies which facilitate engagement within society by asking who the technological society is for – in terms of its relation to the lifeworld – as a first principle of its design, rather than focussing on efficiency, performance, and cost. Feenberg argues that "democratic rationalisation" is necessary to oppose the limits and dogma of technical rationality and industrial capitalism.

He argues that democracy would constitute "an economic and technical requirement of the transition to socialism".[26] He presupposes that "deep democratisation" is inherently a socialist "civilizational project" – rather than a process of decentralisation and localisation – and the transition from capitalism to socialism depends on "an extended period of *democratic struggle over technology and administration*".[27] His argument is premised on the existence of ambiguities in the relations developed through planned action – "margins of manoeuvre" – but he assumes that a single "civilizational project" would be adequate for universal societal development. However, in my view, Feenberg has not adequately explained what is involved in the "deep democratisation" of society, while his vision of a socialist "civilizational project" depends on such an explanation, if it is to provide an adequate vision for the development of a substantive and critical theory of technology. It is not at all clear how we could reconcile technical pluralism with Feenberg's socialist "civilizational project", unless that project was the outcome of the democratisation, rather than its condition. It seems much more plausible that socialism would be one of the means of social organisation available to us, which through democratisation, could be applied in some aspects of societal development (say healthcare or education, for example), while other aspects could be developed through cooperative enterprises or competitive markets. Societal pluralism should allow society to develop without any universal agreement about the "civilizational project", along with all the presumptions about the nature of the good life or what it means to be a human being this agreement would entail. If we accept the truth that we are in a state of innocence regarding knowledge about what the good life is and what is means to be a human being, while simultaneously accepting the value of such knowledge, then we can see how societal pluralism provides society with the diversity and latitude to experimentally explore these questions in a way that does not commit the whole of society to any single guess (which of course, may well be incomplete or an error) and, hence, spreads our bets about how we should try to live a good life and be a human being. The democratisation of the technological society requires a high degree of pluralism in relation to "civilizational projects" if it is to optimise its practical values as the basis for exploring the questions of the good life and what it means to be a human being. It is therefore prudent that society remains tolerant of competing ideas about the choice between civilisations, while encouraging individuals and communities to choose between them or come up with alternatives, in order to maximise the freedom that each citizen has to experiment and explore these deeply existential and ethical questions. The universal answers to societal questions can only be achieved, if they can be achieved at all, as the result of a protracted, democratic exploration of all conceivable alternatives to the point where universal agreement is reached. It is potentially a catastrophic error to establish this consensus artificially as a societal *a priori* in a dogmatic ideology that prevents alternative ideas and practices from being explored. Feenberg argued that there are important distinctions to be made between different technologies and paths of development, but he does not provide us with any rational basis by which such distinctions can be explored and decisions between them can be made, especially due to his assertion of the rationality of a socialist "civilizational project". Contrary to Feenberg's promise, given that he construes "deep democratisation" in terms of his socialist "civilizational project", without any clear account of how "deep democratisation" would take place, it would be open to criticisms from technocrats as being an irrational interference into technical decisions because it requires the normalisation and institutional incorporation of new technical codes into the processes of technical decision-making as an *a priori*, but, given the contradictions between technical codes developed in industrial capitalism and socialism, there is not any technically rational reason to accept it in the first place, within the current technological framework, and the democratisation of the technical sphere would be represented as against the best interest of the public. The possibility of initiating Feenberg's "civilizational project" would depend on a revolutionary overthrow of the current system and the imposition of democratic centralism, which tends towards technocracy and authoritarianism. It is essential that we understand what is involved in "deep democratisation" in order to pre-empt this tendency and to clarify how the technological society could be developed into a democratic society. This will be the topic of the next chapter.

4
Participatory Democracy

History does not provide us with any examples of an ideal democracy. It has been often proposed that the Athenian *Ecclesia* is an historical example of direct democracy, wherein citizens were directly responsible for the administration, legislation, and jurisdiction of the city-state, and had an equal right to speak in the public assembly.[1] By Athenian standards, modern "actually existing democracy" would be quite undemocratic, given that only the elected representatives would seem like citizens to an Athenian, while the rest of the population would seem like slaves who periodically elect citizens.[2] However, in the Athenian city-state, only male citizens were permitted to participate; women and slaves were excluded.[3] The Lombard League of northern Italian towns convened direct democratic assemblies to elect their own administrators, legislators, and magistrates to autonomously govern their own affairs, after they successfully rebelled against the Holy Roman Empire in the twelfth century and forced the Peace of Constance (1183AD), achieving formal recognition of their independence.[4] However, participation in these assemblies was limited to tradesmen and property owning male citizens. Dating from the medieval period, there remain canton and town based direct democratic assemblies in Switzerland, wherein women have achieved political equality, but these small communities tend to suffer from parochialism and social inequalities persist.[5] The Paris Commune of 1871 was a deeply inspiring example of democracy to nineteenth century thinkers, including Proudhon, Bakunin, Kropotkin, and Marx, and has since been deeply influential on theories of participatory democracy.[6] Paris was divided into 60 districts, within which citizens assembled to administrate and legislate their own affairs, appointing delegates (subject to close scrutiny and recall) whenever deemed necessary, until it was brutally destroyed by the French army. Robert Owen's foundation of a factory based on workplace democracy and the New Harmony commune in Scotland in 1825 has been widely taken to be an early and small scale experiment in participatory democracy, and Kibbutzim, first founded in 1909, still exist today in Israel and provide examples of how participatory democracy can work in small agricultural and religious communities.[7] The popular government in Georgia in 1919 (elected with over 80% of the vote) began a democratic revolution in public participation in governance and workers' control in industry and agriculture, including land reforms, a cooperative market economy, and a parliamentary and liberal constitutional legal framework, until it was destroyed by the invasion of the Bolshevik lead Russian Army in the following year.[8] In America, the New England townships and town meetings, some of which date back to the seventeenth century, provide examples of how local citizens continue the democratic tradition of gathering to legislate and administer local affairs.[9] However, these remain small face-to-face meetings to control the local budget and propose town ordinances (wherein levels of participation are inversely proportional to town size) and increased centralisation of administrative and legislative authority in state and federal institutions, which further disempower local communities and citizens, has resulted in decreased attendance in town meetings because they have less practical value. The Montreal Island Citizens Union (in Canada) and the participatory budget in Porto Alegre (in Brazil) have also been cited as contemporary examples of participatory democracy at work.[10] Indeed, Montreal and Porto Alegre are important experiments, but we must also be aware of their limitations, given that, in practice, the federal and state authorities control all but one or two percent of the municipal budget, while centralised federal government and international organisations, such as the World Bank and the International Monetary Fund, are able to effectively override the local budgetary decisions over even this very small percentage. Arguably, the agricultural collectives and industrial cooperatives set up between 1936 and 1937, during the Spanish Civil War, provide the most impressive examples of participatory democracy in action in workplaces and communities, which, even under stark conditions, successfully increased production, improved working conditions, provided social provisions, and even began to modernise factories and farms.[11] The collectives and cooperatives in Spain between 1936 and 1939 show that direct democratisation in the workplace does not necessarily have to be limited to small-scale experiments, but can practically organise large-scale industries and agriculture. They also show that the failure of these experiments was not intrinsic to direct democracy, but was rather due to the incompatibility between authoritarian government and democracy. Of course, such human endeavours were limited and lead to excesses, as many of their detractors are keen to remind us, but, under the conditions of a civil war, while resisting a military coup by the Francoist forces and the efforts of the Stalinists in the Republican government to undermine all factions it did not directly control, in many respects, the collectives and cooperatives showed remarkable ingenuity, morality, and success in developing economic organisation and providing the political conditions for freedom.[12] However, they were short lived. They were destroyed by external violence and suppression, by the Republican Communist Brigades before the Francoist forces delivered the *coup de grace*.[13] We will never know how successful these social experiments would have proved themselves to be. No doubt, such experiments would have been partial successes and failures, even in their own terms, requiring ongoing refinements and modifications to deal with their flaws and limits. Whether in Ancient Athens, the Swiss cantons, the Paris Commune, the Russian soviets of 1905 and 1917, the collectives during the Spanish Civil War, the Hungarian councils of 1956, or the townships of New England, all efforts and experiments in direct democracy have been limited and flawed. Perhaps, we should agree with Rousseau's claim that democracy is the politics of gods rather than mortals.[14] However, we should also be aware that history does not provide us with any example of an ideal society of any kind. This should be unsurprising, given that human beings are limited and flawed. Every kind of political regime, from monarchies to fascism, have been flawed and limited, even in their own terms, and awful and shameful acts have been conducted in the name of the king, the state, law and order, freedom, the revolution, as well as in the name of the people or the divine. The objections against direct democracy based on human fallibility and wickedness are applicable to every kind of political regime and social organisation. Even contemporary "representative democracies" and "republics" (with all their "checks and balances") have been responsible for terrible acts of oppression, corruption, and irrationality. One only need to give the excesses that have occurred in recent years in the name of "democracy promotion" or "national security" a moment's reflection to see that "actually existing democracy" is capable of overriding individual rights, running secret prisons and torture chambers, supporting brutal dictatorships, and unleashing powerful military forces on civilian populations, which has resulted in the death, disability, or displacement of hundreds of thousands of men, women, and children. In comparison, there is not a single historical example of a direct democracy that has led to the level of abuses of power, atrocities, and mass murder achieved by "actually existing democracy".

Since the 1960s, the term "participatory democracy" has been widely used within grass-roots political movements and has a rich history of radical ideas, experiments, and efforts of many activist groups and movements, such as the peace, green, students', women's, gay rights, cooperative, and communitarian movements, as well as in both permaculture and "alternative" culture.[15] It is also a central term in a growing body of academic literature discussing the developments of political critiques of liberal, socialist, and Marxist theoretical traditions describing how democracy could work in industrial societies.[16] Theories of participatory democracy are very much indebted to the nineteenth century anarchist ideas of Proudhon, Bakunin, and Kropotkin – at least to the extent that they tend to share a vision of an industrial society being comprised of federated free-associations between decentralised communities and workplaces organised on libertarian and egalitarian principles of voluntary collectivisation through direct democracy. Rudolf Rocker's vision of an industrial society based on anarcho-syndicalism has inspired theorists of participatory democracy and tackled the question of how democratic ideals, such as freedom of expression and association, workers' self-management, local autonomy, direct action, mutual aid, spontaneity, and libertarianism, could be incorporated into an industrial society.[17] His vision of federated workers' associations controlling the means of production, as a free-association of cooperative labour, to satisfy all the needs of every member of society, was developed from the ideas of Proudhon, Bakunin, and Kropotkin, inspired by the CNT in the Spanish Civil War, and profoundly influential on anarchist theorists of the twentieth century, such as Daniel Guérin, Murray Bookchin, Noam Chomsky, and Colin Ward, as well as radical Marxist thinkers such as Anton Pannekoek, Rosa Luxembourg, and Antonio Gramsci.[18] These thinkers have called for a social revolution to eliminate the State by dissolving all centralised authority, governmental oversight, and bureaucracy by removing all the coercive social and political institutions that suppress the possibility of the creation of the federation of workers' associations as the means to abolish monopolies, emancipate workers from the economic elite, and liberate people from the totalitarian tendencies of the State (in either its state-socialist or capitalist forms). However, participatory democracy has theoretical roots that precede the Industrial Revolution. In William Godwin's study of the problem of governmental coercion, *Enquiry Concerning Political Justice*, published in 1793, we can see an anticipation of the Industrial Revolution, its potential to remove the necessity of labour and liberate human beings from material constraints, and its threat to human freedom.[19] Godwin advocated community-based government by direct democratic participation, wherein national assemblies would only be convened whenever it was necessary to do so to settle common problems, and disputes between communities would be settled by juries of arbitration. In many respects, his resolution to the problem of government coercion and his anticipation of industrialisation amounts to one of the first steps towards a developed theory of participatory democracy in an industrial society. The early twentieth century political theorist G.D. Cole developed Rousseau and Mill's "classical" theories of democracy in order to apply them to industrial societies.[20] Cole was very influential on contemporary advocates of participatory democracy, such as Pateman (who further applied and developed Rousseau's theory) and Macpherson (who further applied and developed Mill's theory).[21]

The thickening of thin democracy

Benjamin Barber's idea of "thickening thin democracy" involved achieving participatory democracy (which he termed "strong democracy") through a gradual process of increasing the level of public participation in the existing institutions of representative democracy (which he termed "thin democracy") rather than dismantling them.[22] This would allow participatory democracy to emerge within a liberal tradition of respect for law, rights, and individual liberty, alongside the public tolerance of differences within a pluralistic, diverse, and open civic society.[23] He also argued that participatory democracy would be a development of the communitarian and "classical" republican traditions, which advocates strong local government and a high level of public commitment to fostering good citizenship, civic virtues, and the development of civic society as a public good.[24] Public commitment cannot be ideologically imposed through either intervention or revolution, on the part of some external agent (such as an invading and occupying army) or any internal intellectual elite (such as a "revolutionary vanguard"), as if seizing power, setting up elections, writing a constitution, and declaring the beginning of a democracy was sufficient for the creation of a democratic society. Democracy is an ongoing process, not a state or particular set of institutional arrangements. It is a civic society defined internally by its citizenry and the decisions they make to constitute themselves as the *demos*. The emergence of democracy, on such an account, is inherently a grass-roots movement, within which the development and differentiation of the form, structure, and content of society are brought together into three aspects of a unity through the democratic participation of the citizenry. It is the outcome of a social evolution, rather than a political revolution. This can only be achieved as the result of an ongoing historical and cultural process of the development and differentiation of how to best understand "democracy" in theory and practice. The possibility of participatory democracy depends on the widespread public education and consciousness about this historical process and the cultural traditions from which the current institutional arrangements have emerged. The potential for democratic participation very much depends on the existence of an educated and committed citizenry, without which, civic society cannot exist at all. In this respect, participatory democracy should not be understood as an abstract, utopian political philosophy, but as giving political form to situated and ordinary acts of civic participation, volunteer work, and helping one's neighbours, as performed by citizens in their local communities. The theoretical task for advocates of participatory democracy is not to provide models for how ordinary citizens should cooperate to develop their communities, workplaces, schools, hospitals, action groups, etc., but, rather, theory should relate such ordinary civic practices to the ideals of participatory democracy and, thereby, connect them with the historical process of the development of democratic participation in both theory and practice. It is this educational aspect of democratic participation that is crucial for the purpose of understanding how the transition from a representative to a participatory democracy could occur in practice.[25] Even though widespread public participation is the crucial foundation for any genuine democracy, the development of the democratic *ethos* requires the prioritisation of universal access to high quality education and impartial protection under the law, which must prescribe the norms and ideals of the vast majority of citizens. By combining the liberal ideals of liberty (or negative freedom) and respect for private property with the "classical" republican ideals of civic responsibility and citizenship, Barber argued that participatory democracy would be

> ...a much less total, less unitary theory of public life than the advocates of ancient republicanism might wish, but it is more complete and positive than contemporary liberalism. It incorporates a Madisonian wariness about actual human nature into a more hopeful, Jeffersonian outlook on human potentialities...[26]

Barber was also inspired by anarchism, especially in its social libertarian form, which promotes the ideals of equality, voluntary cooperation, and positive freedom. However, one important difference between

Barber's strong democracy and social libertarianism is that the latter calls for the revolutionary destruction of the state apparatus, whereas the former is an evolutionary development from "actually existing democracy" and calls for a gradual withering away of the State as more and more of its functions become absorbed into civic society. Participatory democracy would be an evolution of liberal democracy, rather than a radical break from it, but it would transcend the limitations of liberalism by recovering the communitarian and social libertarian aspirations for positive freedom and local autonomy, as a social evolution of human communicative and cooperative capabilities within an open-ended, complex, and changing world. Unlike anarchism, participatory democracy does not require some "year zero" revolutionary event for its possibility, but, instead, would continually develop a participatory civic society in direct proportion to the public willingness and ability for people to administrate, legislate, and magistrate their own affairs. Rather than emerge from some great historical act of mass violence or enlightenment, a participatory democracy involves the continuous development and differentiation of participatory opportunities for societal development. Democratic communities will succeed or fail on the basis of the trust, commitment, respect, and compassion for each other that the members of those communities have and cultivate, as ordinary citizens come together and take personal responsibility for the local administration and development of civic society.[27]

Liberalism assumes that human beings are essentially private, asocial beings that engage in public affairs only to preserve their private interests, and, public administration and legislation requires a level of expertise that is only possessed by the minority of well educated and wealthy citizens or those citizens with specialised, technical training in bureaucratic administration or law. Theories of participatory democracy challenge these assumptions and assert that human beings are both private and social beings, and, the distinction between the private and the public is a product of political activity, not its condition. Specialised training, wealth, or privileged education does not provide or guarantee the acquisition of knowledge of universal truth, morality, or goodness, and, there are not any necessary or sufficient connections between political expertise, rationality, and civic virtue. The relation between political expertise and rationality is open to deliberation and exploration; citizenship and civic virtue would be outcomes of democratic participation. Public participation in the development of civic society (including administrative, legislative, and judicial affairs) would be the educational and transformative means for the development of civic virtue, political efficacy, and social responsibility for the majority of citizens. Historically, the city has been one of the primary loci (alongside the rural commune) for theories of direct democracy.[28] Barber considered neighbourhood and municipal assemblies to be the most important sites of negotiation, wherein citizens and their delegates could deliberate and coordinate the administration and legislation of their local affairs, including healthcare, education, sanitation, transportation, lighting, etc., through the co-operation between other assemblies, local workplaces, and other organisations.[29] Regional, national, and international assemblies would be convened, whenever deemed necessary, through alliances between city based assemblies to achieve common goals and develop economic relationships. A federation of modern cities would be able to perform all the functions of the modern state (including organise mutual defence through trained citizen militia and reserves), without requiring any centralised authority or bureaucracy. Barber argued that neighbourhood assemblies are important for recovering democratic participation and restoring the sovereign power of the citizenry. A neighbourhood assembly should be convened whenever any citizen wishes to raise a matter of concern or suggest a proposal to his or her neighbours. Even though a neighbourhood should have a dedicated space for assembly, an assembly does not need to be formal, it does not need regular hours, and attendance should be voluntary. The citizen should merely inform his or her neighbours that s/he wishes to convene the assembly in advance and invite all interested parties to attend. Given that participatory democracy operates through enrolment, rather than consensus, the purpose of the assembly is to communicate and coordinate voluntary participations, contributions, information, and distribute resources. Decisions would be non-binding in the sense that they would be voluntary agreements between citizens to cooperate to achieve a shared goal. They are statements of intent and plans, rather than decrees and resolutions, and do not involve citizens who either disagreed or did not attend, unless those citizens change their views at a future date. It is in this respect that participatory democracy can be considered to be a libertarian and collectivist form of democracy, within which citizens decide how to organise their own efforts and resources, rather than based on "majority rule", wherein the majority dominates dissenting minorities, once a consensus has been established. Neighbourhood assemblies would be meeting places and mobilisation centres for neighbourhood residents, and, they could be used to communicate and cooperate with other neighbourhood assemblies and enhance each neighbourhood's capacity to administer its own affairs.[30] They would also have educational value for helping citizens to develop their understanding of the meaning of citizenship, civic virtue, social responsibility, and community.[31] For Barber, New England town meetings offer a starting point for how we could envision local democratic assemblies.[32] Despite their limitations, town meetings, as a local branch of legislative government, are capable of dealing with complex local agendas and concerns, while also helping develop civic virtues and skills, such as cooperation, sociability, social responsibility, civility, solidarity, tolerance, restraint, and humility. They also help develop communitarian commitments and values, which despite the risk of parochialism, are fundamental for genuine democracy based on common enterprise and cooperation between citizens.

As Barber argued, although Madison feared democratic excesses, his ideas of the Republic were dependent on the existence of a citizenry that knows how to govern local affairs and take an interest in national affairs. Otherwise, on what basis could citizens be expected to recognise and choose good representatives? On what basis could citizens hold their representatives accountable? Individual self-interest and ambition must be balanced by the cooperative discovery and realisation of common goods, visions, and responsibilities in order for any system of democratic governance to work for the benefit of the vast majority of people. The State is only able to effectively protect the constitutional checks and balances, including the institutional separation of powers and the rights of the individual citizen, if citizens are able to participate in maintaining the political conditions for the preservation of their political freedoms.[33] Liberal critics of participatory democracy have often failed to take this into account when they argue that participatory democracy places too heavy a burden upon the citizenry.[34] One needs to take the realities of preserving the conditions for political freedom into account, if one wishes to preserve political freedom. Politically passive citizens, only concerned with the private realm and the satisfaction of their individual preferences, will not be able to develop the skills needed to preserve their political freedoms and, therefore, will be overly reliant upon a politically active minority to do it for them. Under such circumstances, one should not be surprised when the majority have their freedoms radically curtailed by the minority, acting in their own interest. The administration and legislation of public affairs may well be time consuming, but the level of complexity in modern society requires a high level of attention and participation in order to acquire the skills and political efficacy required to deal with that level of complexity. If citizens are unable or unwilling to exercise the required level of attention and participation then they will be unable to protect their own rights and freedoms. Equal rights will depend on equal participation. Political freedom requires the defence of negative freedom through exercising the positive freedom to participate in the societal development of the institutional arrangements of political and economic relations and activities. It is through democratic participation in securing political freedom (over and above the necessary means to satisfy and protect private interests) that the public realm of human relations can be liberated from the technocratic "nation-wide administration of house-keeping".[35] By emerging from within the liberal tradition, democratic participation will afford pragmatic and progressive opportunities for citizens to improve their confidence and political efficacy, while avoiding many of the possible mistakes and injustices that could occur through "excesses of democracy" while the citizenry are inexperienced in governance, and also provide institutional frameworks and support for the ongoing development of democratic processes and opportunities. The liberal tradition would also provide standards of legitimacy and constitutionality that can ground the legal basis to challenge the continuance of any antidemocratic efforts to undermine or dominate democratic participation. By participating in the administration and legislation of local affairs, citizens will also gain the skills and experiences required to develop political efficacy at a national or international level and make alliances and associations that will further empower their local communities and workplaces.[36] Participatory democracy requires a dynamic and revisable process of interpreting how democratic participation should be conducted, within local contexts, which allows citizens considerable latitude and flexibility in their experiments in how to conduct assemblies at local, region, national, or even international levels. It requires a dynamic and revisable process of interpreting the nature of citizenship and the boundaries between the public and private aspects of human life. Decentralised and localised experiments would gradually emerge as a careful and practical form of societal development. Should any experiment make mistakes, these would be localised and their impact of wider society would be dampened, while providing useful experiences and information for other communities and workplaces as well. By maximising societal pluralism and diversity, there would be a societal stock of alternatives, while citizens would be able to continue any traditional practice that they found to be satisfactory for their local communities or workplaces. The motivation for experimentation would be to discover better ways of organising practical activities, rather than something done for its own sake, and, the decision where and when to experiment must remain a local matter made by the people likely to experience the consequences of such a decision. A strong democracy could be achieved by creating cooperative free-associations between various democratic assemblies which agree to cooperate in accordance with shared respect for the right of each assembly to govern their own affairs.[37]

Participatory democracy advocates the grass-roots collectivisation of political and economic organisations among citizens in order to create alliances capable of effectively resisting and opposing powerful governmental agencies and corporations. By collectively negotiating contracts and the conditions of their participation, without relying on the authority and power of a centralised state apparatus, citizens can confront and challenge organised power, whether corporate or governmental, through the organised power of the citizenry. Participatory democracy radically transforms the conception of the role of the political party. The "traditional" role of the political party is to gather support for a policy platform or agenda, in order to achieve an electoral mandate for administration and legislation. Participatory democracy would transform political parties into educational, persuasive, and consciousness-raising organisations that enrol citizens into collective action to implement and develop particular programmes and projects. The political party would be transformed from being a "representational" to an "ideological" instrument of organisation and co-ordination. Many citizens would have "overlapping membership" of several political parties and multi-party alliances would be commonplace for specific issues, concerns, or proposals. The political form of a participatory democracy is that of a decentralised society wherein citizens have sovereign authority to develop an egalitarian, libertarian, and rational society through democratic assemblies, free-associations, and alliances. Such a society requires a high level of knowledge, skills, and political efficacy among the citizenry in order for local problems to be solved through collective action and political parties should help citizens meet these requirements. Citizens would make decisions through enrolment into cooperative efforts, resulting in a decentralised structure of local-outwards efforts to build networks of participants, through a process of enrolment, rather than either a top-downward hierarchy or a bottom-upward process of consensus formation. In this respect, it would radically differ from competitive elitism and democratic centralism. It is important that democratic participation provides the means for the successful harmonisation of individual goals and social purposes, but the discovery of what constitutes examples of successful harmonisation, as well as deciding the criteria under which we would make such qualifications, needs to remain exploratory and pluralistic – effectively decided in context at a local level – rather than prejudged in terms of theory. The process by which such decisions are tested must remain laissez-faire in relation to other communities and workplaces, rather than decided through centralised planning and coordination, on the basis that, through trial and error, stable alliances and agreements would arise between different workplaces and communities simply on the assumption that it is in their own best interest to achieve them. Theories should be critically developed as heuristics to help us propose experiments in workplace and community democracy, rather than axiomatic models for defining democratic participation, which is something that must remain at stake, discovered and decided through democratic participation. There are also irreducibly aesthetic and contextual aspects of face-to-face communication that are essential for the development of meaningful democratic participation, within local contexts, which cannot be understood in the general terms of theory. Different communities and workplaces will develop democracy in different ways and an over reliance on any theoretical account will result in a definition of democratic participation that will not be universally applicable or meaningful. Even with the best intentions, over-reliance will tend towards ideological dogmatism and its concomitant impractical and coercive democratic centralism. Of course, at a theoretical level, we can specify a universal set of minimum conditions for the possibility of freedom from coercive power and equality of participation, for example: individuals must be able to learn how to develop their diverse qualities and abilities in order to realise their potential; communities must be able to collectively protect themselves; individuals must agree to respect privacy and tolerate differences; freedom of association and freedom of expression must be valued and practised; individuals must have the right to choose the economic conditions of their labour, given their abilities, skills, and available resources; society must provide free and public access to scientific knowledge and technological innovation, as well as open and free access to communications; and every citizen must have the right to participate in any public assembly on any matter of direct concern. However, in order for these to be meaningful as a set of minimum conditions they must be recognised by the vast majority of citizens, and whether they are satisfied in local contexts will be a matter of interpretation.

Whereas there is little doubt among theorists that direct democracy is possible in homogeneous societies, its practical value is open to question, but we face the reverse situation in heterogeneous societies, wherein the practical value of direct democracy is evident, but its possibility is in doubt.[38] Participatory democracy offers the possibility of developing a heterogeneous society into a dynamic polyarchy of allied associations of citizens, forming powerful groups and organisations.[39] Alliances between confederated democratic assemblies, alongside courts, organisations, and networks, would provide the means by which citizens are able to communicate and cooperate with each other, while preventing the possibility of the centralisation of power and authority. It is on this basis that citizens could have the direct sovereign power to administrate, legislate, and magistrate public affairs, without having to be Rousseauesque gods, while our lack of divine foresight or perception is the best reason why we should divide this sovereign power among us in equal measure. This would empower democratic participation and collective action through polyarchic and decentralised networks of free-associations, forming alliances to deal with specific issues and particular projects, without any need for authoritarian and centralised government. A polyarchic citizenry would form an allied majority only for the purpose of satisfying over-lapping interests and concerns. An allied majority would not be forthcoming for extreme or excessive proposals, given that there will be a low level of over-lapping interest in the success of such proposals. It is also highly likely that the polyarchic majority would counter the possible emergence of any dominant faction, except on matters with a high level of common agreement, and a polyarchic majority would fragment in response to extreme or excessive proposals. A polyarchic majority would be incapable of acting in a tyrannical manner because it relies on a high degree of commonality and consent among the citizenry. As Dahl argued, the fear of "mob rule" or "the excesses of democracy" are actually overstated within a modern pluralistic society, given that the majority is actually an equilibrium state of overlapping membership.[40] Furthermore, each citizen has over-lapping membership of several associations, groups, and organisations, which allows them to democratically participate in societal development and political institutions on many different levels and in different capacities.[41] Of course theorists of participatory democracy should be concerned with problems that might occur through "excesses of democracy", such as lynch mobs, pogroms, or other injustices or excesses of zeal, but one must ask, should such abuses occur, whether they are more likely to be stopped and corrected through a decentralised democratic alliance of powerful organisations, or through a massive state apparatus of centralised bureaucratic, police, and military power controlled by a minority in the name of the majority? It is arguably the case that both systems would localise abuses and dampen conflicts in more or less equal measure, but, whereas local abuses may go unchecked for longer in direct democracies they would be unlikely to spread beyond the locality without being opposed, but in an indirect democracy, even though there are checks and balances against the minority abusing its position of power, should it overcome these (through conspiracy, public manipulation, or a *coup d'état*) then it will be extremely difficult for the majority to stop these abuses of power, due to the power of the state apparatus in shielding the minority from the majority through the use of police and military forces.

The risk that a participatory democracy could degenerate into chaos or civil war can be avoided by strengthening the democratic assemblies through cooperative federations and alliances in accordance with their over-lapping interests and concerns. During the process of decentral-ising power and increasing local community power, through federations and alliances, citizens would have sufficient time to learn how to co-operate and administer their own local affairs through reasonable and practical deliberations. The process of gaining power through cooperative action (rather than competition) should teach people how to collec-tively coordinate their actions, without resorting to violence or coercion. The "thickening of thin democracy" is not to be mistaken with reform-ing the state apparatus because, from the onset, it amounts to eroding the State through an ongoing process of deconstructive decentral-isation and constructive participation as citizens learn how to govern their own affairs to their own satisfaction. Withering away the State would be the direct democratic process of gradually making the State obsolete by absorbing its operations into civic society as citizens take over its administrative, legislative, and judicial functions, in that order, as a social evolution from a mass society to a rational, egalitarian, and libertarian society. It is only after citizens have become proficient in administering their own affairs that they should begin to legislate for themselves through citizens' assemblies and democratically write their own constitution. It is only after citizens have become proficient in legislation that they can democratise the judiciary, without risking injustices or excesses of zeal. During this transitional stage, wherein cit-izens take over the administration of society, the liberal constitutional framework of law and the appeal to the judiciary will act as essential guides to avoid abuses of power and excesses of zeal, while citizens are learning how to be good citizens and deepening their commitment to the practical value of participatory democracy by learning how to deliberate and make decisions in the absence of centralised standards or arbitration, and, by critically examining their habitual norms and practices. A participatory democracy would not be a system of local, autocratic "majority rule"; lynch mobs and kangaroo courts do not constitute legitimate expressions of democratic participation, if they contradict the Constitution. By refining and developing the liberal tradition of law and jurisprudence, while citizens are learning how to administer local affairs, a conception of constitutionality acts as a guide to the process of transition, after which citizens can learn how to legislate through democratic assemblies and magistrate through juries. Once the administration of society is conducted through democratic participation, citizens will be in a position to take over the legislative powers of the State as well. The ongoing deliberation and refinement of the Constitution, through democratic assemblies and referenda, will be a crucial component of successfully navigating this transition. It is only once citizens have embodied their own constitutional understanding of the nature of democracy and citizenship (including their rights and duties) in legal practices and conceptions of justice, will they be able to competently transcend the liberal tradition and achieve participatory democracy. The Constitution should be an agreement between all citizens (or at least the vast majority) if it is to act as a General Authority against which the legitimacy of local sovereignty can be challenged. This allows individual citizens to challenge the legitimacy of the legislative and judicial decisions of their neighbours, through appeals to the Constitution, claiming that particular local decisions contradicted the Constitution, as agreed by the same neighbours. If necessary, it is only on the basis of appeals to the Constitution that citizens could enrol the aid of other citizens to defend them, without risking degeneration into factionalism or a civil war. This is essential for the development of political efficacy among an active citizenry capable of learning how to deal with the problems of local self-governance.

By taking the liberal tradition as its point of departure and holding that the General Authority of the Constitution is an essential *educational* aspect of the transition from representative to participatory democracy, by acting as a heuristic focus for how the *demos* assembles and understands itself, the limits of democratic participation can be balanced with the individual expectation of negative freedom. It is only in virtue of having these limits and this expectation that positive freedom can be a part of political activity at all. John Rawls argued that a liberal constitutional protection of pluralism was necessary because

> ...a basic feature of democracy is the fact of reasonable pluralism – the fact that a plurality of conflicting reasonable comprehensive doctrines,

religious, moral, and philosophical, is the normal result of its culture of free institutions. Citizens realise that they cannot reach agreement or even approach mutual understanding on the basis of their irreconcilable comprehensive doctrines.[42]

If public institutions embodied the comprehensive doctrines of some citizens rather than others, then any cooperative enterprise deliberated and decided through such institutions would tend to endorse these doctrines as positive freedoms, while also treating them as negative freedoms and, thereby, preventing others from challenging them. Hence, Rawls argued that public reasons for mutually binding forms of social organisation and cooperation should be decoupled from comprehensive doctrines and appeal only to "free-standing political values", as explicitly stated in the constitution. The institutional structures of deliberation and decision-making in the public realm should only be based on "free-standing political values", shared by all citizens, while the employment of comprehensive doctrines should be confined to civic society and the private realm. However, this is all well and good as an abstraction, but how do we decide what "free-standing political values" are? The "free-standing political values" that Rawls endorses in the US Constitution have emerged from a cultural background and entail comprehensive doctrines regarding conditions for human freedom, equality, and happiness, even if we all agree that freedom, equality, and happiness are goods.[43] Given that "free-standing political values" entail comprehensive doctrines in defining the boundaries of constitutional protections of pluralism, it is essential that the citizenry (or at least the vast majority) are involved in collectively deliberating and deciding what these values should be. The citizenry must be able to deliberate and agree upon the Constitution, which must remain open to re-evaluation and revision, and in this sense "free-standing political values" are the political values agreed by the vast majority, if not all. Otherwise the Constitution will assert the comprehensive doctrines of a minority as "free-standing political values", which will either distort public reason, by only permitting the outcomes that benefit that minority at the expensive of the majority, or it will simply become an historical document without practical value. It is in this sense that participatory democracy becomes a condition for the preservation of liberal pluralism. By recognising that every comprehensive doctrine is equally worthy of consideration and it should be left to particular citizens in particular circumstances to decide for themselves which comprehensive doctrine they wish to try to live by, theories of participatory democracy are able to embrace liberal pluralism at a societal level, preserving negative freedom through the societal commitment to a principle of localisation, while also embracing the practical value of particular comprehensive doctrines at the local level as an expression of positive freedom through a principle of free-association. In this respect, the tension between the local and societal levels will constitute its own check and balance against the centralisation of power and authority. A participatory democracy would satisfy Galston's requirements for constitutional liberal pluralism that the demands placed on the citizens are not too heavy for them to bear; the "free standing political values" are not too diverse to exist in the same society; and its broad outlines must correspond to the moral sentiments and circumstances of members of that society.[44]

Once we take Galston's requirements into account, alongside the recognition that Rawl's "free-standing political values" entail comprehensive doctrines, then we can answer Dahl's claim that there are some universally accepted values, rights, and liberties that override the democratic process and majority consensus.[45] Such values, rights, and liberties could only be universally accepted through the democratic process and majority consensus. In my view, any such universal values, rights, and liberties would by discovered through the process of discovering what the democratic process means in context and these values could only maintain their universality due to the commitment of the vast majority to their reasonableness (as universals) on the basis that it is a condition of having such individual rights that the individual respects the individual rights of others. Without the agreement of the vast majority, it would be impossible that any set of rights could be universally respected. Instead they would be based on the imposition of assertions and privileges of a minority over the majority. Any such overriding values, rights, and liberties would be based on coercive power relations. In Dahl's example of "a fair trial", he claimed that it is only the individual's right to due legal process that overrides the desire of the majority to punish whomever is accused of some vile crime, but he failed to recognise that any such act of "mob justice" would also require the suspension of democratic values, such as free speech (of the defendant and the accusers) and free association (having skilled representatives to aid the defendant and the accusers), which would involve the suspension of the democratic process, as well failing to satisfy the practical need to actually protect the public from the guilty party by correctly identifying them. If we possessed absolute foresight or universal epistemological principles of jurisprudence then we would have no need of the democratic process at all, but, in their absence, the democratic process is the best means to "a fair trial". Of course, even with all due diligence, sometimes the democratic process will make mistakes, such as convicting the wrong person for a crime, but this is due to human fallibility, rather than a failure of the democratic process. Human fallibility is the reason why the democratic process has practical value in the first instance and the individual's right to "a fair trial" inherently involves the democratic process, as the best means to realise the public good (correctly identifying criminals and protecting the public from them), and, if the majority did not respect this right (by stringing up the first suspect from the nearest tree) then this would not be an outcome of the democratic process, but, rather, a failure on the part of the majority to engage in the democratic process. A lynch mob cannot be considered as an outcome of a democratic process in the same way that a cattle stampede cannot be considered as an act of democratic agreement formed by cows to decide which direction they should run. In this sense, once we consider the practical value of the democratic process as the means to realise the public good, we can recognise that "majority consensus" is necessary but not sufficient for a decision to be democratic, if that consensus has been achieved without a process of public deliberation directed towards discovering the best course of action. Instances of "mob rule", lynch mobs, or the exclusion of minorities would be incompatible with participatory democracy because they would be the consequences of failing to engage in democratic processes of deliberation and decision-making among those concerned with the consequences of any course of action.

Democratic assembly and reasoned deliberation

Pateman held that theories of participatory democracy were indebted to Rousseau's philosophical conception of genuine democracy, even though Rousseau also held it to be an impossible ideal for human beings.[46] Rousseau's democratic philosophy was a theory of how citizens could empower and liberate each other by coming together and participating in the administration and legislation of public affairs. Through the use of reason, each citizen would come to discover for themselves that participating in the formation and execution of "the General Will" was a condition for freedom and power. This was not a statement of conformity to some collective consensus, but was a statement of conformity to the outcomes of the shared human capacity to reason. Reason allowed one to become free from the dictates of the passions and the arbitrary whims of other human beings. The validity of Rousseau's political philosophy stands or falls on whether any such shared human capacity to reason actually exists or not. According to Rousseau, the use of reason allowed human beings to freely come together and work out how to cooperate in order to achieve shared goals that were unavailable to naturally free individuals acting alone. It is only on this basis that the abdication of natural freedom and the conformity to reason was justifiable.[47] The common good could not be realised simply by individuals conforming to the majority decision. It can only be discovered through the process of using reason to formulate "the General Will", as a cooperative outcome of reasoned deliberation, to the intellectual satisfaction of all concerned individuals. The reasoned discovery of "the General Will" is done freely because the individual loses nothing and gains everything by conforming to the outcomes of reasoning to the satisfaction of their own intellectual and moral conscience. Rousseau advocated a Socratic sense of citizenship wherein intellectual virtue is a condition for civic virtue, which is a condition for happiness and political freedom. According to Rousseau, all human beings, as naturally free individuals, wish to live a free and happy life, a good life, but the conditions for living such a life are not subjective nor created by dictate. These conditions must be discovered through reasoned deliberation about experience and expectations in a world that does not conform to human desires or dictates. When individuals coexist, all seeking to live a good life, reasoned deliberation becomes the best means to decide the best course of action, without any citizen being subjected to the arbitrary wishes or whims of any other. The citizen exchanges their natural independence (secured only through natural power) for civic rights (protected as inviolable under the law), which obtain their compulsion because the individual has freely agreed to their reasonableness. The laws are the social contract obeyed by the same citizens who made them acting in the capacity as the sovereign power. In this respect, it is the capacity of the citizenry to act as the legislature that is fundamental to a genuine democracy. Participation in reasoned deliberation has a profound transformative effect over and above simply performing a calculation of the sum of expressions of subjective preferences and assertions. Reasoned deliberation has a transformative psychological effect, by transforming how we understand ourselves and our relations with each other, and it also has a transformative ontological effect, by transforming how we understand our potentiality and limitations within the world.

Citizens should submit to the sovereign authority of "the General Will" only when it has emerged as an outcome of the reasoned deliberation between all citizens. In this respect, Rousseau was neither authoritarian nor proposing some kind of democratic totalitarianism. For Rousseau, "the General Will" is a general authority in the sense that, through reasoned deliberation, it is the authority that all citizens agree to abide by. This is distinct from the specific authority of doctrines and dictates forced upon all by either a minority or a majority. It is not the most popular preference, but is that which every person can agree upon after reasoned deliberation. It is in the self-interest of any individual to voluntary cooperate with others and equally share the labour and benefits of any cooperative endeavour with others. The political freedom remains, according to Rousseau, for the individual to choose between the natural freedom of the individual to be independent and the desirability of common goods that can only be satisfied through reasoned, cooperative endeavours. Individuals will not benefit if they do not cooperate because they will not be able to achieve specific goals without cooperating. In this respect, the individual citizen conforms to the law because it is in the interest of all citizens if every citizen does so, and, by freely following his or her own reason each and every citizen becomes interdependent with every other citizen, while being independent of the subjective will or dictates of any one citizen. Laws should only arise as the outcome of reasoned deliberation between all citizens, in the sense that the authority of law arises from the reasoned agreement between all citizens, rather than the coercive power of any one citizen to impose their subjective will or dictates.[48] It is not simply the case that each citizen agrees to abide by the majority decision, regardless of how the majority makes that decision. The process of deciding "the General Will" is a participatory process of discovering which laws all citizens (or the vast majority) can agree to abide by, on the basis that abiding by these laws would be in the interest of all citizens (or the vast majority), and, therefore, the law is an expression of the satisfactory outcome of reasoned deliberation between all citizens. By conforming to the law, the individual citizen satisfies their own self-interest by abdicating their natural freedom when it contradicts the outcomes of reasoned deliberation. Freedom involves the master of one's life through exercising one's capacity to reason, discovering the conditions of the good life, rather than the ability to exercise absolute power over others or the unrestrained acquisition of luxuries. It is important to note that, for Rousseau, "the General Will" was the reasoned agreement of all citizens, not an aggregate sum of individual wills or the outcome of majority vote. It is only when circumstances require a rapid decision, would it be practical for the vast majority to act in accordance with the consensus, but this would be a substitute for "the General Will", rather than an approximation. It would be a temporary measure, until "the General Will" could be determined through the reasoned agreement between all citizens. One's conformity to "the General Will" is literally an act of commitment to the best course of action for the individual in relation to the best course of action for any individual in the same circumstances. By freely committing oneself to the outcomes of reasoned deliberation, one becomes liberated from arbitrary authorities and dogma, in order to freely pursue the good life, as a human being capable of exercising reason to discover the conditions for truth, goodness, and justice. The social contract is the shared commitment to the law and the human capacity to reason is a condition for political freedom. The threat to political freedom comes from inequality of participation in reasoned deliberation or the suspension of reason altogether when deciding laws binding on all citizens. It is the failure to commit to and realise the human capacity to reason that is the threat to political freedom. On this account, the simple-minded process of tallying the sum total of individual preferences and opinions, quantifying already established, habitual prejudices and assumptions, either through elections or opinion polls, would not constitute the democratic process at all because it does not involve reasoned deliberation between citizens in a public assembly, wherein any and every citizen could freely participate and exercise their capacity to reason to the satisfaction of their intellectual and moral conscience.

For Rousseau, the idea of democracy was taken to be governance without government. In a democracy the affairs of the State would be performed by all the citizenry, in accordance with laws decided and enforced by the citizenry. Given that all laws would be those that all the citizens considered to be the best possible laws for each and every citizen to live by, the fullest realisation of freedom could only be achieved by obeying the law because these laws are those that apply to all citizens and have the good of each citizen as their end, therefore, each and every citizen would freely choose to agree when each and every citizen uses their capacity for reason. The capacity to reason is the condition for political freedom. Rousseau's theory of democracy requires that all citizens are equally capable of exercising reason.[49] Inequality of education would undermine the possibility of democracy, if those inequalities resulted in some citizens being unable to exercise their reason. Rousseau also considered economic equality to be a condition for political equality, but he did not require it in an absolute sense. Some economic inequality is permissible, providing that it is relatively slight and it does not lead to some citizens being beholden on others for their livelihood. As long as all citizens owned their housing, have access to the economic means of sustaining their existence, and slavery of any kind was prohibited, then differences in wealth would not translate into political inequality. Providing that associations between individuals are diverse and (approximately) equal in economic and political power, then these associations will not threaten the freedoms of others. Democratic participation is an educational process within which private interests and preferences are informed and transformed by the public process of discovering public goods and deciding how best to achieve them through deliberative reason and cooperative action. Practising democratic participation teaches individuals how to exercise their reason to negotiate and cooperate with others; it also transforms the individual conception of what goals are desirable and how to realise them. Reasoned deliberation has a psychologically positive effect in transforming a private person into a public citizen, allowing the individual to explore the potentiality and limitations of social being at a deeper and more meaningful level. It has existential and ontological importance. It also provides individuals with the ability to act collectively as equals to resist the selfish demands and dictates of powerful individuals, while also compelling equally placed individuals to learn how to take into account the interests of others to gain their cooperation through persuasion. On this account, individual intellectual and moral conscience is essential for the judgement of the best course of action. By discovering how to reasonably and practically cooperate with others in order to discover and realise shared goods, one liberates one's life from the whims and dictates of arbitrary authorities and conventions by intellectually engaging with the conditions, meaning, and consequences of one's actions within a society. It is the ontological commitment to discovering and living the good life, in the absence of certain knowledge, which compels the individual to participate in a democratic process of reasoned deliberation and decision-making that is directed towards discovering and realising the public good. The democratic process becomes the consciously driven intellectual process of cooperatively unfolding the goodness of social being in a changing, complex, and open-ended world. Communication is fundamental to Rousseau's conception of either republican or democratic forms of civic society. The limits of communication place limits on the size of any genuine democracy, and, even though Rousseau denied that any such democracy had existed and he was sceptical that it was even possible for mortals, he considered the city-state (which would be a small city by modern standards) to be the theoretical upper limit for a democratic society.

Direct democracy must be based on face to face communication in local assemblies to maintain a close relation between individual participation and the outcome of deliberations, which is essential for developing social responsibility and political efficacy.[50] Democracy depends on the participation of citizens in a public assembly to deliberate and decide public affairs, and, as Arendt argued, political speech can only be properly realised through the public realm, distinct from the administration of labour and economic activities. It is political speech that allows human beings to fully realise their potential as human beings and makes human life worthwhile.[51] It is only in the public realm that citizens can *come together as equals*, wherein each contribution can be considered as equally worthy of consideration by their peers.[52] To be human is to be an equal participant in the public realm as a citizen within a community, with equal rights and duties to each other, as determined through political speech. To be deprived of access to the public realm is to be denied the possibility of being fully human. The political life is inherently worthwhile – a good life – in the sense that it is good to live it and to be deprived of it would be harmful. For Arendt, political speech within the assembly of equals is an opportunity for human beings to bring out the best in themselves, discover who they are, as well as determine the best course of action. It brings forth an ideal of citizenship based on radical individuality and a shared humanity, which discloses a shared world, by bringing together plural and diverse perspectives; the meaning of the public realm is provided through the coexistence of different people who experience a shared world differently because they experience it from different positions.[53] The public realm has existential significance, as well as practical value, wherein human beings can discover their identity and potential within a shared world.[54] By providing the conditions for the freedom of expression, interpretation, criticism, and deliberation of different perspectives on truth, justice, and goodness, it is the *res publica* wherein human beings can discover for themselves what is best in them. The public realm is the site for human freedom, when realised through political speech. If political freedom means anything at all it must mean participating in the governance of one's own affairs – participating in the public pursuit of freedom and happiness as the ground for fully realising what it is to be a human being. Ideals such as freedom, equality, and justice are products of political activity, and, consequently, unless the whole of society participates in the deliberation and decision-making processes that give these ideals practical meaning in context, then these ideals will be understood and implemented in a way that suits their producers and not necessarily in a way that is good for the rest of society. It is essential that these remain contested terms. Following on from Arendt, Barber argued that, even in the absence of universally agreed epistemological and moral standards, deliberative and decision-making processes, made necessary by the absence of universal standards, can be guided by maxims, rules, and principles, providing that these are open to change through deliberation.[55] These guides are abridgements of traditions and prior deliberations. Providing that citizens are aware of their provisional and historically contingent nature, these guides can be used as a means to help the democratic assembly deal with uncertainty, to make decisions about the nature of public life, about how to deliberate and decide which courses of action are likely to promote the public good, and how the public good is to be discovered and understood. In this respect, the task of a genuine democracy is to invent itself, as an applied, provisional, and flexible political *praxis* – an ongoing creative and chosen path of developing an understanding of a shared world. The norms produced "by an ongoing process of democratic talk, deliberation, judgement, and action" are legitimised "solely by that process, which exhibits and refracts the political culture's changing circumstances and evolving communal purposes" and, consequently, they are provisional and conditional upon the open and flexible achievement of common modes of discourse and vision.[56] The legitimacy of any political judgment is not based on establishing how it can be logically derived from some set of moral or epistemological standards, but, rather, through its practical value in aiding a genuinely democratic assembly to define itself and clarify its shared vision for the future. Understanding political talk is crucial for making the deliberative process intelligible. Barber noted that it involves understanding

...its creativity, its variety, its openness and flexibility, its inventiveness, its capacity for discovery, its complexity, its eloquence, its potential for empathy and affective expression, and its deeply paradoxical (some would say dialectical) character that display's man's full nature as a purposive, interdependent, and active being.[57]

Political talk allows the discovery of shared goods and visions; the articulation of interests, bargaining, negotiation, and exchange; persuasion; agenda setting; exploring mutuality and commonality; affiliation and affection; defending one's rights and the rights of others; bearing witness and self-expression; critically developing and refining understanding and plans; and community building. The practical value of democratic participation can only be discovered through social inclusion within a heterogeneous, pluralistic, and diverse political talk. The discovery of this practical value is possible if and only if different communities can use and communicate their knowledge, ideas, experiences, and concerns, through cooperative free-association, in order to discover and realise shared visions of the good life, human well-being, and the ideal society. As Arendt argued in *On Revolution*, Jefferson presupposed that all human beings have a natural right to establish new societies in accordance with their reasoned understanding of which laws and regulations are most likely to promote public happiness. This necessarily involved the right for human beings to actively participate in the governance of their own affairs. Neither freedom nor happiness can be understood as simply a subjective or private matter, in this respect. That is not to say that privacy is not a part of the enjoyment of freedom or happiness, or that someone cannot discover freedom or happiness through private pursuits, but it means that the conditions for freedom and happiness, for human beings, involves an agreement with others, both in terms of discovering the conditions of freedom and happiness in communion with others and also being aware that one's own pursuit of freedom and happiness has consequences for others. While it is evident that there is not any universal agreement about the content or meaning of freedom or happiness, it is the case that one learns their content and meaning from the public realm, and that each and every human being's ideas of freedom and happiness are as worthy of consideration as anyone else's. Even if we agree that human beings should be free to understand happiness in whatever way they please, either privately or publicly, when such an understanding underwrites any proposed course of action then there are likely to be public consequences of that understanding and, therefore, once reason and agreement form the basis for the public deliberation of public consequences, the freedom or happiness of one is dependent upon the freedom and happiness of others. Government can only hope to provide the means or conditions by which citizens can pursue freedom or happiness if those that govern are likely to suffer the consequences of their decisions. Thus government must ultimately be legislated by the people, if it is to be for the people; otherwise it would be inevitable that government that allows only an individual (king or despot) or a privileged class to monopolise the public realm will be premised on the use or threat of violence in order to serve only the interests of one or a minority at the expense (or independently) of the interests of the majority. In this respect, if the aim of revolution is political freedom, then it must be premised upon the liberation from the use or threat of violence, and must maintain the conditions by which people remain free from coercive power and oppression. Hence, a constitution should not only limit the scope and powers of government, but it should also express the spirit of political freedom that a people wish to take upon themselves to exercise.[58] It must be a product of ongoing reasoned deliberation if it is to correspond to the changing public understanding of the nature of citizenship, law, rights, duties, and the purpose and limits of government. Such a public deliberation will not only express changes in how citizens view their rights and duties *qua* citizens, but also clarify the ongoing discovery of the nature and conditions of freedom and happiness. In this sense, a democracy is a permanent revolution based upon how human beings understand themselves and the world. It is important to note that Madison's emphasis on the need for a republican government to protect the rights of individuals or minorities from the majority is only possible when a majority constituted by the over-lapping membership of minorities and individuals combines in order to agree upon this limit to the legitimacy of the majority. It is a self-imposed limit by the majority as an expression of its political freedom. Hence, as Arendt argued,

> ...power comes into being only if and when men join themselves together for the purpose of action, and will disappear when, for whatever reason, they disperse and dissert one another.[59]

Workers' control and cooperatives

Theories of worker control, as developed by Cole, Rocker, Pannekoek, Luxembourg, and Gramsci, advocated the workers' direct democratic control over the organisation and conditions of labour, through workers' assemblies (or councils). Management and administration would be the responsibility of elected managers and boards of directors whose decisions were to be ratified by the assembly (or council). Workers would control the conditions of their own labour and the operations of their work, but the means of production would be socially owned property. The Yugoslav ideas of "market socialism" and the system of communes and workers self-management enterprises were theoretically modelled on ideas taken from these theories of workers' control.[60] However, the Yugoslav system was deeply flawed in practice and had very little relation to participatory democracy. This was largely due to the constant interference by the Yugoslav Communist Party, which imposed directives; controlled the nomination of candidates for election to workers' councils; controlled trade union leadership; fixed prices, taxation, and wage levels; imposed bureaucratic controls over enterprise planning and coordination; forced successful enterprises to subsidise unsuccessful ones; imposed production targets and priorities; aggressively punished all political dissenters; controlled all media; and suppressed criticism of the workers' management system and the "market socialism" model.[61] Even though the productivity of the Yugoslav system was comparable with similarly sized European capitalist economies, such as Spain or Italy, even until the late 1980s, while being more equitable, and, it was much more decentralised than comparable communist countries, such as Cuba and the Eastern European regimes, while, again, being more equitable, the Yugoslav model resulted in an economic system that was troubled by a lack of "internal discipline" in the workplace, due to an absence of motivation or accountability; a high level of nepotism and corruption; the political appointment of unqualified Communist Party officials or supporters to technical or managerial positions; excessive regulation and bureaucratic demands on elected managers; managerial initiative unrewarded if successful, but punished if not; carefully developed enterprise proposals often ignored and replaced with arbitrary Communist Party directives; and an absence of any real democratic participation in economic or technological development. In many respects, the Mondragón cooperatives in the Basque region of Spain are better examples of workers' control in practice, due to being also examples of worker ownership, and show how cooperatives offer a genuine alternative to both industrial capitalism and state-socialism.[62] Studies show that successful cooperatives have organised training and educational programmes, disseminating important technical knowledge and skills throughout the membership, which have provided economic benefits alongside helping members develop their social skills and overcome social inequalities.[63] Cooperatives operating in a competitive market tend to reproduce many of the practices of a capitalist enterprise, but, given that workers are also owners, with a stake in the success of the enterprise, they also alleviate many of the pernicious aspects of industrial capitalism, such as capital flight, outsourcing, and union busting, which have resulted in the decline of real wages, loss of benefits, and increased insecurity of employment.[64] A cooperative market economy would be a mixed economy in the sense that it would allow laissez-faire competition to test producer decisions, establishing states of market equilibrium through negotiation with suppliers and consumers, but, as a cooperative system, within which all workers are also owners, it dissolves the labour market because the exchange value and surplus value of labour would be identical. Cooperatives may appoint professional managers, but due to the workers' control over management, cooperatives also avoid many of the problems that arise due to hierarchical managerial structures.[65] Providing that cooperatives are based on workers' control, equitable profit share, cooperative ownership of the means of production, and individual ownership of one's labour, a cooperative market economy could provide the economic basis for decentralised, rational, sustainable, and libertarian development of society.[66] This would provide the economic conditions for greater political equality and a broader education in democratic participation within wider society.[67] Cooperative enterprises can operate as "shelter organisations" that provide educational, financial, and technical support to other cooperatives within large networks of cooperative enterprises.[68] Scientific research and development could be incorporated into a cooperative market economy through publicly funded "shelter organisations", which could include universities and research institutions, without requiring any centralised control or administration.

Workers' control involves more than providing the means to improve productivity and workers' conditions. It provides the conditions for political freedom to emerge from the democratisation of the relations of production in accordance with both the technological imperative and humanistic concerns (such as egalitarianism, social justice, volunteerism, etc.), which educate workers in democratic participation and integrate workplaces and local communities, without requiring centralised planning or coordination. Openness and trust between workers, as well as genuine opportunities for each and every worker to participate in the decision-making process, are necessary to achieve the degree of solidarity and the level of commitment to democratic participation required for any system of workers' control to be an effective and practical alternative to hierarchical systems of management.[69] As Harold Lydall put it,

> The productivity of human beings is not simply a matter of the equipment available to them, nor of the amount of formal education or training which they have received, but also of something else – certain e*spirit de corps*, attention to quality and detail, responsiveness to demands from outside the enterprise and from managers and workers within the enterprise.[70]

This *espirit de corps* can only be achieved if a workplace is also a community, wherein each member of that community has a stake in its development. One of the fundamental problems with the so-called Marxist-Leninist communist systems, such as the Soviet Union and

Cuba, is that nationalisation of industries and the centralised control over the economy prevents individuals from having any stake in the outcome of their efforts. The Communist Party doctrine of attempting to equalise income levels between workers in any given industry or service, regardless of the levels of productivity or efficiency of any particular enterprise, while making surpluses or deficits a matter of socialised property or risk respectively, undermines "workplace discipline" because it disassociates reward from effort. The efforts on the part of the Communist Party to induce a voluntary work ethic by appeals to "the revolution" and "moral obligation" were disastrous failures as the basis for economic development.[71] Due to the State control over the decision-making process and the methods for compensating for differences through the distribution of collectivised surpluses between enterprises within the same industry or service, it proved quite impossible to maintain high levels of voluntary work discipline, often under harsh conditions, when workers did not have any personal incentive to improve productivity and efficiency by taking personal responsibility and risks, working harder, and experimenting with new techniques. As a result, the Communist Party increasingly resorted to authoritarian and coercive decrees and methods, while, the systems of redistribution and allocation became increasingly prone to corruption, nepotism, and bottlenecks. In a cooperative market economy, mixed kinds of ownership are possible, but it is essential that workers own the product of their labour (i.e. receive equitable profit share). The surplus value of produce could be decided through market exchanges between local producers' and consumers' cooperatives, but the exchange value of labour must be equal to its surplus value. In a cooperative market it would be impossible for anyone to profit from someone else's labour, but workers *qua* owners would have a direct stake in the cooperative enterprise and, therefore, voluntarily subject to "workplace discipline". Once individuals have a personal stake in any cooperative enterprise, they are increasingly likely to participate in collective endeavours and decision-making processes. It is essential that workers' democratically control the conditions and organisation of their labour, but this does not prohibit workers from electing or appointing managers or directors. Workers' control over cooperative enterprises promises increased productivity due to beneficial psychological and sociological effects of control over working conditions alongside the motivational effects of profit share; increase in efficiency due to reduction of waste, repetition, lost working hours (due to sickness or industrial disputes), etc; reduction of costs due to the elimination of superfluous middle management and supervisors;

reduction in administrative paperwork, human resource management costs, and conflicts with unions; increase in quality of products and services due to "market discipline" (when motivated by profit share); increase in personal initiative and creativity once workers have an equitable stake in the enterprise; increases the chance of radical innovations and refinements of the enterprise's operations; increased benefits to local communities, greater integration of workplaces and the wider community, increase in level of social responsibility and civic virtue; tendency towards economic equality within any given local community; improved security of employment, local economies dependent only on local cooperative market conditions, which decreases local vulnerability to global market fluctuations and the threat of capital flight; and, cooperative credit unions and exchanges would tend towards economic equality between communities, without needing a centralised redistributive bureaucracy.

Large-scale industries could be democratically organised through federated trade unions, wherein workers would control their respective trade union via elected delegates to an assembly (or council) of federated unions, which would provide a forum for communication and coordination, but delegates should not have decision-making powers and should be subject to recall. Federated trade unions would provide a network for cooperation between workplaces, as decided by the workers, to form regional or industrial alliances between trade unions, without requiring a centralised bureaucracy or a hierarchical system of management. Such a system can maintain "workplace discipline" and "market discipline" through voluntary participation and fair trade, while also providing workers and their families with healthcare, education, and other social benefits. Initiatives, experiments, and direct action would not require centralised approval, allowing each workplace and union autonomy, but other workplaces and unions would be able to deliberate these and react as they deemed appropriate. Production, distribution, and consumption would be based on mutual agreements, contracts, and exchanges between producers, distributors, and consumers, without needing any centralised planning or controls. Furthermore, a cooperative market economy would tend towards market localisation, which would result in the tendency for greater economic stability and security within local communities. It would also dissolve the distinction between workplace and community, making both aspects of the public realm and rejecting the assumption that economic activity (work and consumption) is a private matter, which would dissolve the distinction between producers and consumers, especially when social needs are served by local markets. This would tend towards overcoming conflicts of interest between citizens *qua* workers and citizens *qua* residents, allowing greater democratic control over local economic development, overcoming workplace alienation, and offering an effective counter to either technocratic administration of community development or leaving it to the caprice of "market forces". A cooperative market economy would allow decentralised economic and community development, while also integrating production and consumption, at a local level, leading to rational and sustainable economic activities. A cooperative market economy would act as a levelling mechanism on the extremes of wealth generated by industrial capitalism and globalisation, but would greatly alleviate poverty and alienation for large numbers of people. The tendency would be towards communities with a high degree of economic equality between moderately comfortable citizens, each capable of satisfying their needs and changing the conditions of their existence through democratic participation, rather than a society of extreme wealth/power and poverty/alienation. The assumption that hierarchical and centralised organisations are necessary to operate complex technological systems is based on little more than conformity to the fact that once the *status quo* has been established then it is very difficult to change it and develop an alternative. It is usually the case that such centralised and hierarchical organisations have been able to use the power and resources to secure a monopoly in order to perpetuate the public perception that this organisation is necessary and its organisational structure is the best way to control and direct the system. However, it is not the case. As Richard Sclove has pointed out, the San Francisco Bay public transit services are managed by over thirty independent organisations, without any overall system of centralised control, and is able to provide trolley cars, buses, rail, subway, ferries, car pools, and satisfy the transportation needs of a large and heterogeneous urban area.[72] Postal services have also been able to operate at local, national, and international levels, without any overall system of centralised control. There is not any good reason to assume that technological systems, such as water purification and distribution, electric power generation, agriculture, telecommunications, media, etc., require hierarchical and centralised systems of organisation. Through legally elected public representation at policy and managerial levels, it is possible to make any organisation more accountable and transparent, and through increased public participation in the strategic and tactical development of such organisations it is possible to make them more pluralistic and integrated within the communities they serve. The greater degree of elected public representation at all levels within any large organisation would lead to greater degree of public participation in the everyday operation and development of that organisation. In any genuinely democratic society, it would be pluralistic opportunities for cooperation that would determine the degree of political freedom that people have in the way that their activities are organised and structured to achieve public goods. Once cooperative endeavours are afforded genuine opportunities for practical experimentation, as well as learning from the experiences of others, within a pluralistic and diverse society, a mixed economy would be the inevitable result, allowing different communities and workplaces latitude to further experiment and cooperate with others in various forms of ownership of different enterprises and industries. By decentralising social organisation there would be considerably greater scope to explore alternative paths of scientific research and technological innovation, while also further developing and coordinating communication networks through economic federations and alliances. Democratic control over the means of production, distribution, and transportation would bring together the conception of community and workplace, directing the organisation of local knowledge, experience, intellect, and character in order to develop productive capacities of local workplaces to realise a shared local vision of community life and public goods, while deepening social relations and the sense of fellowship between workers as a community within a wider community. Research and development of "the forces of production" and various forms of "relations of production" could take the form of a strategy or plan achieved through democratic consensus, as an ideal, but, assuming that there will usually be an absence of universal agreement, the democratic process should remain a decentralised process of discovery and experimentation based on principles of free-association and localisation. Deliberating and deciding the ends and means of work should be a dialectical, political activity, achieving social agreements that have causal effects upon the organisation of labour, creating new possibilities and potentials, allowing the discovery of new goals and intentions, which can be achieved through cooperation. By examining the organisation of labour and goals as a fundamental aspect of the ontology of social being for the development of our communities and workplaces, we can see how economic and political activity would converge within a participatory democracy to be dual aspects of decentralised democratic processes of social evolution.

5
Science and Technology

It is essential for theories of participatory democracy to take into account technical expertise in order to develop a plausible vision of how direct democracy could develop within a modern society.[1] Theories of participatory democracy need to embrace and incorporate technical and scientific expertise into their discussion of the democratic process in the absence of any universally agreed epistemological or moral standards. Science and technical expertise are important for the possibility of participatory democracy, providing that they are considered to be subordinate to intellectual conscience and democratic values, directed in accordance with the aspiration for everyone to be able to discover and realise the good life. As such, science and technology are important aspects of the recovery of the potentiality for the evolution of society into a rational, egalitarian, libertarian, and sustainable society, just as they are important aspects of the potentiality for the degeneration of society into irrational totalitarianism or barbarity. Everything is at stake, providing that the majority awaken from their conformist slumber to recover their freedom to cooperate to realise the potential for human enlightenment and emancipation. It is impossible for any centralised authority or technocracy to predict which skills or knowledge are going to be valuable or important in the future or how knowledge or skills should be distributed throughout society. People should be able to choose to develop whatever skills and knowledge they consider important or valuable, and, even though it is of practical value to society that people develop diverse skills and knowledge, it should be left to citizens to decide their educational or training paths for themselves. Hence, it is essential to decentralise the deliberative and decision-making process to allow people to choose for themselves which path they wish to take; dividing their resources in direct proportion to the shared commitment to the path. However, it is increasingly difficult for citizens outside the scientific community to evaluate the judgements of scientists, especially in the context of a history of sustained strategies to exclude the public. This makes it even more difficult for citizens to trust scientists when citizens are alienated from the scientific tradition, its ideals, and its processes of achieving consensus. Darin Barney claimed that there is a crucial antagonism between democracy and technology because

> ...democracy does not require substantial expertise as a qualification for participation in decision making, and so it allows for government by mass ignorance; technology, as it becomes increasingly complex, requires for its control and development levels of expertise that exceed the capacity of most citizens and, thus, it defies democratic governance.[2]

However, Barney assumed that it is possible to achieve levels of expertise capable of controlling and developing technology. It may well be the case that it is quite impossible for most people to judge the complexities and implications involved in the research and development of nanotechnology, biotechnology, medical science, space exploration, computer science, and nuclear fusion, but, as the current scientific debate about "global warming" has shown, with all the disagreements between scientists regarding the consequences of burning fossil fuels and deforestation, it seems quite unlikely that scientists will be able to judge the complexities and implications of these brave new technosciences either. There is not any "intellectual elite" that is capable of governing the construction of the technological society. Of course we are simply not in a position to ignore scientific knowledge, but we should not rely on the foresight of scientists either. We need to critically and publicly question the rationality and desirability of the societal goods provided by science and technology. This involves critically examining our goals and ideals from broad and diverse perspectives, opening public reflection and deliberation to neglected aspects and criteria, and, given the absence of universally agreed moral and epistemological standards (even among scientists), this process requires democratic participation.

Informed participation is a condition for effective participation.[3] Do citizens have a sufficient degree of scientific literacy to meaningfully participate in scientific research and technological development? It is obvious that a low level of scientific literacy among the general public is a serious obstacle to democratic participation in scientific research and technological development, but, this problem also extends to scientists outside their area of specialisation, as well as concerning broader levels of education in the humanities (including the history of science and technology) and a philosophical understanding of the public good.[4] The high degree of narrowly specialised and technical education among scientists is also of great significance as a limit to democratic participation when scientists decide the criteria of public debate about the directions of science and technology. Elected professional politicians and legislators are often as technically illiterate and poorly educated in the sciences as most other citizens. Hence, they routinely delegate political decisions about the implementation and development of the technosciences to unelected officials, technical experts, bureaucrats, and committees. Putting aside all the possible corrupting influences exerted by contractors upon unelected officials (as well as the influence of lobbyists on professional politicians), any officials, technoscientists, and politicians are no more impartial or better placed than any other citizen to decide what is the best course of action, and many of the decisions about the policies for the implementation and development of the technosciences are not necessarily made with the public good as being an important consideration. And, even when these decisions are made for the public good, these "experts" may well have conflicting conceptions with the general public about what the public good is. As a consequence of this, it is imperative that the public has greater involvement in the whole decision-making process, especially when contracts are allocated, and the whole political system is more transparent and accountable. However, increasing representation is only a temporary measure that prepares the way for greater public participation in the whole political system at every level. This would mean that citizens would have a greater guiding influence on the direction of the implementation and development of the technosciences and there would be a reasonable confidence that governance reflected the plurality and diversity of the citizens who, due to more pressing interests elsewhere, did not participate in those particular decisions. A genuine democracy does not prohibit a division of labour regarding the political decision-making process, but it does prevent decisions being made behind the scenes by unelected officials and vested interest groups who consider themselves to have some kind of hegemony over governance. Of course it will be helpful that citizens have a better education in the technosciences, as well as a broader education in politics and citizenship, but what is essential is that there is greater local democratic governance of the technosciences, the removal of technological and political obstacles to democracy, and that there are legal protections of the individual and democratic rights of citizens. Of course improved communication between citizens and scientists would be helpful for better informing citizens about science and allowing citizens to question scientists about their work, but this is still a one-way communication – the citizen is being "informed". In order for genuine communication between citizens and scientists to be actually achieved, citizens need to have opportunities to "inform" scientists about their concerns and the broader societal criteria that need to be addressed. Genuine communication offers the chance that scientists and citizens can better understand the meaning of science for societal development by cooperating with each other. As Kleinman argued, "lay citizens" are capable of learning the complexities of scientific methods and technical procedures but, it is not simply the case that the public understanding of science involves the improvement of levels of scientific literacy, in the narrow sense of learning scientific and technical discourse.[5] There needs to be a broader public understanding of science among scientists and lay citizens alike. Citizens need to be involved more at the preparatory and design stages, while scientists need to learn that there are broader concerns and issues at stake than a set of technically specialised problems and their solution. The challenge is to optimise participation, as an ongoing process, in the harmonious integration of scientific research and technological innovation into already existent structures of the world, in order to increase the potential for stability and understanding. Once we broaden the meaning of scientific literacy and recognise that democratic participation needs the time and resources required for careful deliberation and decision-making, as well as opportunities to address the assumptions and structures which reinforce social inequality, then we have good reasons to believe that citizens are capable of learning how to effectively communicate their concerns to scientists, as well as understanding the complexities of science and technology. We also have good reasons to believe that scientists are capable of learning how to effectively communicate their concerns to their fellow citizens, as well as better understanding the complexities of the implementation and development of science and technology in the wider world outside of the laboratory or a computer simulator, through a broader philosophical, historical, sociological, and political understanding of science and technology. Rather than providing an *a priori* basis for the exclusion of the public, the complexities of science and technology actually provides good reasons for the inclusion of the public in deliberating and deciding the directions of scientific research and technological innovation.

It is important to address the profound role that science has in the construction and development of society and the everyday lifeworld. As Richard Sclove pointed out, the general unreflective and uncritical social complicity in decisions regarding the implementation and development of technologies

> that haphazardly uproot established ways of life is as perplexing as discovering a family that shared its home with a temperamental elephant, and yet never discussed – somehow did not even notice – the beast's pervasive influence on every facet of their lives. It is even as though everyone in a nation were to gather together nightly in their dreams – assemble solemnly in a glistening moonlight glade – and there debate and ratify a new constitution. Awakening afterward with no memory of what had passed, they nonetheless mysteriously comply with the nocturnally revolutionised document in every word and letter. Such a world, in which unconscious collective actions govern waking reality, is the world that now exists. It is the modern technological world that we have all helped create.[6]

Sclove argued that the societal ineptness at guiding technological change is not the result of the complexity of modern technology, but, rather, is the consequence of a failure to democratically evolve institutions and constitutional mechanisms through which we could ask, debate, and act upon appropriate questions (as decided by the public rather than only by technical experts or professional politicians). He argued that this failure has allowed technology to become implicated in perpetuating antidemocratic power relations and eroding social contexts and opportunities for democratic participation. As well as raising questions regarding technical feasibility and utility (in terms of its impact on the economy, environment, and public health) of any proposed technological change, we also should be asking about its social dimensions and impacts (including its political, cultural, sociological, and psychological effects). Moreover, this questioning should not be limited to the effects of the technology in isolation, but it should also focus on the combined effects that emerge from a complex of coexisting technologies. Sclove argued that whilst we continue to neglect the social dimensions of implementing and developing new technologies then we remain unable to address and change our deepest social and personal problems. I agree with Sclove that the antidemocratic power relations in our society are not caused by technology *per se* and profound social transformations are necessary if we wish to live in a democratic society, but, in my view, it is crucial to examine the way that technology, once implemented and institutionalised, develops an autonomy that can resist and erode our ability and motivation to make changes in social systems and relations. Hence, transforming society and technology needs to occur simultaneously. Of course, society has deep-seated social pathologies that transcend technology, such as greed, sadism, fear, ignorance, elitism, racism, sexism, ageism, arrogance, alienation, and religious intolerance, but we need to examine how the technical structures and processes of integration of technologies can position and empower one group in society, while it can marginalise and disempower others. We need to examine how these inequalities and prejudices are inbuilt into, empowered, fortified, and normalised by those technical structures and processes of integration. As Sclove pointed out, being technological is part of being human and we cannot do away with all technology without ceasing to be human, but we should not merely adapt and conform to whatever technologies are innovated. We need to develop adequate approaches and procedures to questioning technology and eliciting alternative technologies more compatible with the kind of society within which we wish to live. Unless we examine the way that inequalities are structured and institutionalised within technologies then we may well find that our technologies do not adapt and conform to whatever good intentions that we may have. It was for this reason that Sclove argued that we need to examine technologies as important species of social structures.[7] If we accept that our actions and identity are shaped by social structures such as family, religion, politics, and economics, then we must also accept that they are shaped by technologies. When our activities produce cumulative material and social results that are not completely determined by social structures, these are fed back into and transform our society's evolving structural complex. Sclove termed this process as *structuration*.[8] He argued that this explanatory concept should be used in a normative argument for proposing that the procedures for implementing and further developing technologies should be guided by a principle of democratic structuration. For Sclove, the democratic structuration of technology is a moral responsibility of the highest order, founded on our respect for moral freedom as being a fundamental human good and right, and is a necessary and substantive constituent of a modern democratic society. Of course, using evocative terms such as "moral freedom" invites considerable philosophical reflection and debate. Such a debate is beyond the scope of this book. However, while I have considerable philosophical sympathy to the idea that democratic participation is a moral good and fundamental human right, it is not the purpose of this book to argue for this idea. My argument is that democratic structuration has practical value for the rational development and implementation of scientific research and technological innovation because it optimises social pluralism and, therefore, increases the social capacity to intelligently and creatively adapt and respond to events in our messy, complicated, and capricious world. Participatory democracy enhances society's capacity to incorporate and utilise social diversity, which increases our chances of developing sustainable practices and enhancing the longevity of humanity by rationally constructing an ecologically sustainable and technologically advanced society in accordance with egalitarian and libertarian principles, given an absence of universally agreed epistemological and moral standards in a complex, changing, and open-ended world.

Science, faith, and society

Michael Polanyi's philosophy of science, as described in his book *Science, Faith, and Society*, has the rare distinction of being a realist philosophy that acknowledges that science is based on little more than a faith in the social conventions and traditional practices of the scientific community.[9] Polanyi argued that all realist claims about the objectivity of scientific knowledge of natural laws are based upon a personal commitment to a cultural belief in the existence of natural laws as a real feature of Nature that exists beyond human control, independently of scientific knowledge of them, and these laws cause an indeterminate range of consequences, some of which will be unknowable and unthinkable. The faith in the social conventions and traditional practices of the scientific community presupposes the belief that it is possible to discover natural laws through scientific activity and the meaning of all scientific research and training depends upon this shared belief in an underlying causal structure to reality.[10] Without this belief, it would be impossible to sustain the idea that scientific knowledge of "the general nature of things" is universally applicable to explain the experiences of all human beings, in similar circumstances, and corresponds to the underlying realty that is independent from human experience. Assumptions about the underlying causal structure of reality are implicated in the scientific understanding of which questions are reasonable and interesting, what would constitute evidence for the verification or refutation of any possible answer to these questions, and what kind of concepts and relations should be applied to human experience in order to signify perception of the "tokens of reality". Scientific propositions are concerned with disclosing this reality, suggesting possible new experiences from which "the process of their discovery must involve an intuitive perception of the real structure of natural phenomena."[11] However, this belief cannot be verified or refuted in relation to experience, given the available plurality of possible inferences, assumptions, or explanations that are required to relate any given experience to theoretical terms. There is nothing inherent to the experience of observing the movement of planets in the sky that confirms Copernicus' theory and it would be rejected if viewed from a strictly empiricist stance, nor is there anything in the experience of observing an object falling to the ground that confirms Newton's theory of universal gravitation. Concepts and fundamental representations are brought to particular experiences in order to unify these experiences as experiences of an underlying force, mechanism, or causal structure.[12] Polanyi rejected the notion that descriptive exactness, predictive accuracy, or any unifying methodological operation has any role in the epistemology of science over and above the role played by ordinary perception and a scientifically trained intuitive grasp of the underlying causal structure of reality.

Polanyi's epistemology of science was based on the psychological process of intuitively discerning aspects of reality that are not controlled by the observer but are involved in the shaping of perception. How do scientists develop this intuitive capacity? Assumptions regarding the underlying structure of realty are tacitly learnt during scientific training and education, by imitating the discursive and material practices of established scientists, and these constitute heuristic guides to action, rather than the epistemological basis for demarcation between science and non-science.[13] By learning scientific methods and the standards of scientific investigation, through imitating established scientists, these heuristic guides shape scientific intuition, interpretation, and perception. The decision regarding the selection and assimilation of evidence and facts are ultimately matters of personal judgement of the scientists involved in scientific inquiry and practical activity. For Polanyi, it is through the psychological and philosophical study of perception that the progressive nature of science can be understood. Hence he held the view that

> The scientist's task is not to observe any allegedly correct procedure but to get the right results. He has to establish contact, by whatever means, with the hidden reality of which he is predicating. His conscience must therefore give its ultimate assent always from a sense of having established that contact. And he will accept therefore the

duty of committing himself on the strength of evidence which can, admittedly, never be complete; and trust that such a gamble, when based on the dictates of his scientific conscience, is in fact his complete function and his proper chance of making a contribution to science.[14]

Hence Polanyi's philosophy of science differed from Karl Popper's *Logic of Scientific Discovery* in many crucial respects.[15] Polanyi argued that the history of science shows innumerable examples where scientific propositions were not falsified by conflicting observations, but instead suggested a new mechanism to account for the discrepancy, such as Galileo's use of friction to explain the discrepancy between the motion of a ball on a plane and his theory of motion. Polanyi rejected the notion that there is any logic of scientific discovery at all, but instead argued that it is an art, transmitted by examples of the practice that embodies it, without any precisely defined methodology, and, ultimately is based on personal judgement and conscience in being faithful to the scientific tradition. Learning how to practice science involves accepting this tradition and becoming a representative of it. The possibility of science – as an ongoing activity directed towards the discovery of objective truth – depends on the faith that the scientific tradition and its methods are progressive. Hence,

To understand science is to penetrate to the reality described by science; it represents an intuition of reality, for which the established practice and doctrine of science serve as clues. Apprenticeship in science may be regarded as a much simplified repetition of the whole series of discoveries by which the existing body of science was originally established.[16]

At most, principles, such as falsification, act as heuristic guides to action, rather than epistemological principles. Scientific discovery is a tacit process of making decisions and personal judgements, involving intuition, creativity, and heuristics, balanced by critical restraint, learned from within a scientific tradition of historically developed and refined training and practice, without any clear epistemological understanding of how discovery occurs. Science is not a purely empirical pursuit either. Based on the evidence there is no reason to prefer the Copernican system over the Ptolemaic, given that both systems have about the same level of descriptive exactness and predictive accuracy. Both the interpretation of the facts and the facts themselves are based on historically conditioned and contingent interpretations involving general assumptions regarding the requirements of naturalistic explanations in general and also particular assumptions to explain particular observations or experiments. The scientific process of discovery is the culmination of participatory and communicative acts directed towards understanding and demonstrating truth about the causal structures of reality over and above any instrumental value or political expediency. Science is an inherently social process of communication, critically articulating and demonstrating its own truths to the satisfaction of the consciences of the members of the scientific community, in order to provide an intelligible understanding or representation of reality. Polanyi emphasised the role of creativity in this and argued that

> The propositions of science thus appear in the nature of guesses. They are founded on the assumptions of science concerning the structure of the universe and on the evidence of observations collected by the methods of science; they are subject to a process of verification in the light of future observations according to the rules of science; but their conjectural character remains inherent in them...discovery, far from representing any definite mental operation, is an extremely delicate and personal art which can be but little assisted by formulated precepts... All the efforts of the discoverer are but preparations for the main event of discovery, which eventually takes place – if at all – by a process of spontaneous mental organisation uncontrolled by conscious effort.[17]

Choosing the naturalistic interpretation depends on a devotion to science as the best means of discovering objective truth – it depends upon a faith in the rationality and freedom of science. The choice between a naturalistic and magical interpretation rests upon a choice between traditions. Hence, for Polanyi, both scientific and mystical truths are based on faith which can only be upheld from within a community and the only difference between science and mysticism, on Polanyi's account, is between the types of community within which each approach is practiced. Whereas the truth of science cannot be demonstrated to someone who does not share a devotion to science, the devotion to science can become known in terms of the value of the free society it creates within which the continued pursuit of a open and free intellectual process, where individuals are able to openly and freely interpret science in accordance with their intellectual conscience and subject their interpretation to the scrutiny of their fellows. Polanyi's faith in science was premised upon his faith in the value of the open society. However, this still leaves us with the question of how we could philosophically justify the claim that scientific intuition corresponds to the underlying structure of reality? Polanyi asserted that the only explanation for this possibility is that we possess a "faculty to guess the nature of things in the outer world".[18]

Like Thomas Kuhn, Polanyi identified the disciplinary framework of science in terms of the standards and norms inherent to the scientists' values embodied in periodicals and books, the priorities of research institutes and funding bodies, as well as university science departments.[19] Authority in science – "a hierarchy of influence" as Polanyi termed it – is more attached to persons (exemplars) than it is to offices or institutions.[20] It is established in terms of reputation, experience, and expertise, and the unity of science is dependent on scientists knowing who are the experts in neighbouring fields, so to speak, rather than the maintenance of the same minimum epistemological standards in all the fields of scientific activity. On this account, science is itself a social network of experts and their areas of activity, established through evaluations of credibility, trustworthiness, and respect between scientists.[21] Membership of this network is not established in relation to some centralised institution, nor is it established by conforming to some transcendent epistemological doctrine, but, rather, is a local decision that is made by already established scientists about who is considered to be a colleague, a fellow scientist, and a member of the scientific community. In this sense, the decision about whether any activity or proposition is scientific or not is a decision made by already established scientists regarding what is of use or interest to them as scientists, rather than whether it conforms to any abstract epistemological standards. Scientific authority is dispersed throughout the network – the scientific community – and the student submits to the authority of the exemplary scientists whom are taken to embody the scientific tradition in virtue of their credibility, trustworthiness, and the respect that others have for their expertise, knowledge, and skills. For Polanyi, a fundamental aspect of the scientific tradition is its demand that each generation is to critically reinterpret the nature of this tradition in order to better represent it. Dissent and criticism are major aspects of the scientific tradition, alongside intellectual virtues such as honesty, discipline, independence, and originality. By embodying this tradition through training and education, the student learns how to develop personal judgements, rather than rely on appeals to the authority of others, and, once the tradition has been fully embodied, the student can reject authority, assume full responsibility before their own conscience, and become a scientist in their own right. It is by embodying the scientific tradition that the student learns how to dissent, while also utilising the traditional practices and interpretations, and the student learns how to competently criticise scientific authority, while also maintaining a firm conviction in the soundness of the scientific tradition. It is this shared conviction that not only unites scientists in their faith in science, founding their relationships upon trust in each others' shared commitment to the same intellectual virtues, but actually forms the basis for their dissent from consensus. When scientists dissent from the current consensus, they actually appeal to the scientific tradition in order to convince other scientists that they are right to dissent because they hold it to be true that their dissent is more in line with the tradition than is the consensus. It is in this sense that the heuristic premises of the scientific tradition are normative rather than epistemological in so far as they are intellectual virtues and ideals cultivated by the scientific community, as a matter of personal judgement according to individual conscience, rather than unifying methodological principles. Science is in a state of permanent revolution in the sense that the *status quo* is constantly being challenged for its deviations from the scientific tradition and being called upon to restore itself as each generation of scientists applies their personal judgement and conscience to the task of reinterpreting and renewing the scientific tradition. In this way, each generation of scientists challenges the current consensus about how the scientific tradition is to be respected by showing the scientific community how they ought to respect it better.

When a government asserts the premises of inquiry, taking upon itself the decision about what constitutes moral or scientific truths, as well as the responsibility to conduct that inquiry and disseminate its results to the public, either in the form of information or legislation, then, even when it does so in the name of the public good, totalitarianism is the inevitable result. Hence, Polanyi argued that

> Whether a free nation endures, and in what form it survives, must ultimately rest with the outcome of individual decisions made in as much faith and insight as may be everyone's share. Any power authorised to overrule these decisions would of necessity destroy this freedom.[22]

Similarly when big business (multinational corporations) control the aims of science, subordinating the directions of scientific research and the dissemination of its results in accordance with their profitability and marketability, the intellectual freedom upon which science depends is suppressed and science will be destroyed. It is necessary for science that it is a participatory and democratic process within a scientific community committed to open and free discussion between scientists sharing a common devotion, capacity, and obligation to pursue, discover, and communicate scientific truth; that members of the scientific community preserve their spirit of independence, exercise critical reason and dissent from the consensus when and where their conscience dictates that there are ways to better conform to the scientific tradition; and the scientific tradition preserves the intellectual freedom of conscience to decide how to best interpret that tradition. In many respects, the scientific community constitutes a participatory democracy, wherein the democratic virtues are also intellectual virtues. For Polanyi, science constitutes Rousseau's ideal democratic community within which

> Every succeeding generation is sovereign in reinterpreting the tradition of science... [that guarantees] ...the independence of its active members in the service of values jointly upheld and mutually enforced by all.[23]

Every generation of scientists takes on the responsibility of reinterpreting the scientific tradition, in order to best serve that tradition, and should they neglect to attend to that responsibility then science would itself become meaningless as a human pursuit over and above achieving instrumental power or satisfying political ambitions. Polanyi distinguished between "the General Authority" of precepts and prepositions and "the Specific Authority" of doctrine and conclusions.[24] The former is essential for the establishment of common standards and norms upon which science depends, whereas the latter would destroy science completely. The former leaves the decisions for interpreting scientific tradition to numerous individual scientists within a community of scientists dedicated to science, whereas the latter centralises the decisions into a hierarchy that acts on behalf of the community, to which all individuals are to conform to as law. The General Authority is akin to Rousseau's concept of "the General Will" formed through the reasoned and conscientious commitment of individuals to the scientific tradition, as the best expression of their intellectual efforts and aspirations as scientists. The scientific tradition is the basis for the social contract between scientists, wherein each individual scientists has the intellectual freedom to interpret how best to uphold the scientific tradition in accordance with their own intellectual conscience. By making such an interpretation, relying on personal judgement and individual conscience, each scientist takes upon themselves the sovereign power to shape the substance of scientific activity and affect the interests of fellow scientists. When scientists take upon themselves this sovereign power, based upon a shared commitment to science as a whole, the scientists are able to formulate the General Authority, as "the General Will", which does not need any Specific Authority to arbitrate over these individual decisions. The social contract within the scientific community is that of a commitment of all scientists to the ideals and standards of the scientific tradition – a devotion to science as a whole – but, under the General Authority of the scientific tradition, all scientists have the freedom to interpret how the ideals and standards of that tradition ought to be respected, in accordance with their training and personal judgement on matters of intellectual conscience, and, in this regard, whether and how the scientific tradition is to be respected becomes the responsibility of each generation of scientists.[25] Hence, Polanyi argued that

> [It] is impossible to safeguard against the mistakes of such decisions, because any authority established for such a purpose would destroy science. It is in the nature of science that it can live only if individual scientists are regarded as competent to state their views and the consensus of their opinions is regarded as competent to decide all questions for science as a whole... Their decisions are inherently sovereign because it is in the nature of science that no authority is conceivable which could competently overrule their verdict.[26]

However, argued Polanyi, the competence of each generation of scientists to make such decisions for the whole of science is in direct proportion to their conscientious commitment to science as being the best means to discover objective truth because

> ...if we believe in science, we will accept competent scientific opinion as on the whole valid, even though the final validation of any proposition will always involve a fractional amount of personal responsibility on our part.[27]

As Polanyi pointed out, this does not provide any reason for anyone outside the scientific community to share that commitment. This awareness is important because, as Polanyi argued, it is essential that discussion about the truth, meaning, and value of science must remain open and free within wider society, just as how to best interpret the scientific tradition should remain open and free within the scientific community. It is essential for the health of science that it remains open to challenge from rival and alternative interpretations of Nature. How should the outcome of such challenges be decided? Polanyi appealed to the intuition and conscience of citizens in a free and open society, guided by the principles of free discussion, fairness, and tolerance, transmitted by a tradition of civic liberties and embodied in the institutions and practices of democracy.[28] The dissent and criticism of a "judicious public with a quick ear for insincerity of argument is therefore an essential partner in the practice of free controversy".[29] This, of course, depends on the existence of diversity and pluralism in media; the existence and activities of heterogeneous political, cultural, scientific, and humanitarian organisations; the constitutional embodiment of fairness and tolerance in law and custom; and, the public dedication to ideals, such as truth, justice, and charity. The "democratic spirit" of any people is dependent on their respect for intellectual freedom of conscience, their commitment to truth and honesty, the belief that truth can be learned and conveyed, and the common value and shared practice of communication and education. Playfulness, cooperativeness, commitment, and trustworthiness are as important for scientific research as they are for democratic participation. They are the conditions for the open society, which is itself the condition for the democratisation of science and technology.

However, as I argued in *On the Metaphysics of Experimental Physics* and *Modern Science and the Capriciousness of Nature*, scientific truth is dominated by instrumental reason. Even when scientific knowledge does not have any immediate practical application, scientific propositions are "tested" by their consequences for the refinement and development of the instrumentality of the technological activities involved in calculation, measurement, and experimentation. Objectivity is itself understood in terms of the exercise of instrumental reason and the progressive aspect of scientific discovery is inherently bound by its consequences for technological innovation. Hence, even though scientific researchers may well consider themselves to be working in the context of pure research and discovery, their work is emergent from and situated within an historically conditioned and contingent technological framework bounded by human interests and expectations, which are themselves dialectically developed and differentiated as technological innovation transcends prior limitations and brings new possibilities into the world. Human interests and expectations are embedded into the organisation of the trajectories of scientific research and it is in this sense that scientific methodologies can be meaningfully said to be value-laden, even when scientists are genuinely attempting to be disinterested, impartial, and objective. In the experimental sciences, scientific discovery occurs within a technological framework of mapping out the contours of the interactions between human interventions and machine performances. It is within this framework, represented and understood in terms of the ongoing and stratified stabilisation of machine performances, juxtaposed with their theoretical representation in terms of the realisation of natural mechanisms in material arrangements, that connects the intuitions, personal judgements, and technological activities of individual scientists with the historically developed conventions, standards, and expectations of the scientific community. Indeed, modern science is an inherently realist pursuit, which presupposes mechanical realism to explain how the artificial activities of experimental science can discover natural mechanisms, but this is only an internal rationale. From its onset in the sixteenth century, the objectivity of science has been related to its practical value, the truth status of scientific propositions are deferred until tested within the technological framework, and it is implicated in the societal gamble in the rationality of the construction of a technological society as an improvement over the natural world. Hence, although I agree with Polanyi's claim that the personal judgements of scientists are implicated in the conviction and faithfulness of the ideas of the scientific process, in my view, once we take the technological framework into account, locating its epistemological warrant in the precepts of mechanical realism, and also examine the way that scientific experimentation involves controlling and manipulating natural phenomena as the means of achieving certainty and power to mould and pacify a recalcitrant and capricious Nature, as the test for the validity of scientific propositions, we can see how the personal judgements of individual scientists have been historically conditioned and culturally mediated to intuitively understand the natural world in technological terms. Through education and training, the individual scientist learns how to understand his or her connection with both the scientific tradition and the underlying structure of reality in terms of his or her ability to identify, represent, and manipulate natural mechanisms through placing material practice under theoretical description. The scientist's personal intuition at having correctly established a connection with underlying reality through theoretically understood material practice is itself a manifestation of the wider societal gamble in constructing a technological society that is itself the rational realisation of natural laws and mechanisms in technological activity. Critics of science and technology must address the rationality of the societal gamble if they are to seriously challenge the foundations of modern society. The need for a radical change in outlook regarding how we develop science and technology is not only a matter of moral respect for the natural world, but, also may well be a matter of survival for human beings as a species (as well as the many other species we threaten to take with us), as well as examining how the construction of the technological society has become the means by which a minority of human beings can dominate the vast majority by controlling the means by which human beings can sustain and change the conditions of their existence.

The democratisation of scientific research and technological development requires the democratisation of the wider society within which science and technology emerge as historical and sociological phenomena. This involves the democratisation of communities and workplaces, wherein the development and implementation of scientific research and technological innovation can be democratised. This requires local citizens to develop democratic communities and workplace assemblies to coordinate collective action to actively participate in the administration of local affairs and to retake control over local development, while also developing a democratic economic alternative through producer and consumer cooperatives, as well as enhancing their capacity for collective bargaining within unions. Through alliances between democratic assemblies and free-associations, local citizens would be able to take control over local government through the ballot box. Once citizens have taken control over local budgeting and taxation, as well as public tendering and local regulation, it will become possible to directly fund universities and hospitals at a municipal or regional level, which would both allow and require considerable public participation in the direction of scientific research and technological innovation within these institutions, while also promising the possibility of the liberation of scientists from the constraints and limitations of funding through private commerce or centralised government. Universities and hospitals would be able to cooperate and coordinate with other universities and hospitals, forming free-associations, which would liberate scientists from the limitations of "market discipline" and the ideology of privatisation of all scientific and medical research. It promises to provide a socialised infrastructure, without requiring centralised authority and controls, wherein scientists could participate in a growing network of scientists, funded through the public purse, and coordinate their efforts through free-association. The polyarchic relations between networks of scientists – forming a scientific community – and the wider citizenry (of which each scientist is also a member) would be based on democratic processes of developing mutual understanding, exploring overlapping concerns and interests, and negotiating the coordination of efforts and resources through universities and hospitals. This does not prevent scientists from working in private institutions or businesses, but, rather, offers a socialised alternative to the privatisation of scientific research and development, thereby increasing the heterogeneity of science and technology, and, also improving the capacity for criticality and diversity between scientists, who will be working under different conditions, in accordance with different criteria, and have different interpretations of how best to serve the scientific tradition. Even if we agree on the rationality of the societal gamble and the objectivity of science, we have to consider the integration and development of science, technology, and society in broader political and philosophical terms, subordinating science and technology to a democratic process of discovering the public good and the conditions of human freedom. After all, even if the human freedom and the rational development of society consist in the discovery and implementation of natural laws, it does not mean that society should be governed by scientists. In *God and the State*, Bakunin argued that if the administration and legislation of society was placed in the hands of an academy comprised of the most learned scientists, inspired only by the love of truth, who would frame all laws only in accordance with scientific knowledge, this would lead to the most monstrous society.[30] He argued that science is imperfect and assuming that the majority of people do not understand it, otherwise there would be not any need for the academy to frame the laws, the imposition of these laws upon the ignorant majority by the learned minority would result in a mass society of passive and irrational citizens venerating and conforming to dictates based on an imperfect scientific understanding of natural law, *as if it were absolutely true*, without any comprehension of those laws or how they were framed. This would result in a society of mindless brutes, which, by granting scientists absolute authority, would corrupt the critical and revolutionary nature of science itself. Such a society would tend towards an authoritarian, technocratic society wherein the scientists would have to resort to propaganda to instruct citizens, while having to base their own position within the academy in relation to power and the specific authority of whatever doctrine had be imposed on the citizenry, rather than a critical regard for truth and how to best interpret the scientific tradition. If scientists are to attempt to change public policy they must do so *as equals* with their fellow citizens through persuasion and discussion, where citizens come to learn about the science involved and see its truth for themselves, while scientists learn the concerns of the fellow citizens and also develop a broader understanding of the relationship between science and society. In that way, all citizens would develop a rational and active relationship with science, allowing them to develop a critical regard for truth and the scientific tradition also, and, therefore, through reasoned deliberation about science and its implications, they would be better placed to decide how to further develop it, understand what its limits and priorities should be, and guide the implementation of its discoveries in society. This holds true for all kinds of technical expertise as well. As Bakunin put it,

> In the matter of boots, I refer to the authority of [the cobbler]; concerning houses, canals, or railways, I consult that of an architect or engineer. For such or such special knowledge I apply to such or such a *savant*. But I allow neither [the cobbler] nor the architect nor the *savant* to impose their authority upon me. I listen to them freely and with all the respect merited by their intelligence, their character, their knowledge, reserving always my incontestable right of criticism and censure. I do not content myself with consulting a single authority in any special branch; I consult several; I compare their opinions, and choose that which seems to me the soundest. But I recognise no infallible authority, even in special questions; consequently, whatever respect I have for the honesty and the sincerity of such or such an individual, I have no absolute faith in any person. Such a faith would be fatal to my reason, to my liberty, and even to the success of my undertakings; it would immediately transform me into a stupid slave, an instrument of the will and interests of others.[31]

For Bakunin, science should retain its universal and abstract character, but it should be subordinate to considerations of the particularities and practicalities of life, and scientific knowledge and education should be distributed throughout the population.[32]

Scientific research and development as public property

The WTO Agreement on Trade Related Aspects of Intellectual Property Rights (TRIPS) has imposed global standards of patent and copyright protections for the benefit of multinational corporations, which negatively effect the development of poor countries by restricting their access to affordable pharmaceuticals and copyrighted materials, and also disrupt the international free flow and exchange of scientific research and technological innovations.[33] A detailed discussion of the complexities of TRIPS is beyond the scope of this book, but it is important for any theory of participatory democracy to address the issue of patents and intellectual property rights. I shall limit my discussion to "the free-rider problem": the use of scientific knowledge and technological innovations without contributing to the research and development costs. TRIPS was designed to provide international standards and regulations that protect patents and intellectual property rights from "free-riders" and, thereby, preserve the motivation for private investment in research and development. However, TRIPS is based on the assumption that private investors and the market are the main providers of resources for research and development. This assumption is questionable and largely based on a neoliberal ideological faith in the "invisible hand" of the market. In fact, taxpayers have provided the funding to educate scientists and cover the high risk and investment stages of most advanced technologies, including microelectronics, computing, and telecommunications, as well as educating the public to the benefits of such technologies, subsidising distribution, and providing public access.[34] It is also the case that so-called "free-riders" provide economic benefits by disseminating knowledge, increasing competition, reducing costs, and preventing monopoly control over knowledge (including the suppression of patents). It is in the public interest to tolerate "free-riders", providing that the public also accepts the burden of research and development investment costs by publicly organising and funding science and technology. If scientific knowledge and technological innovation are public goods and risks, then it is in the public interest to take responsibility for their research and development, rather than relying on private enterprise, which, understandably, demands profits and ownership when it has to bear the burden of investment. All knowledge in the public realm has public effects, and, therefore, the public has a reasonable claim to the right to legislate and administrate public effects, which includes the dissemination and application of knowledge and innovation. This public commitment will allow the public to benefit from market competition between "free-riding" producers, while also being able to subject scientific research and technological development to public scrutiny by the scientific community and democratic assemblies. After all, if we look at the problem from the perspective of economic self-interest, why should the public accept the burden of the costs of policing and regulating systems to protect private intellectual property rights, with the concomitant protection of private monopoly control over supply and pricing, as well as paying for the private research and development costs figured into the market price of the product? It would be more cost effective for the public to directly fund research and development at public universities and research institutes and then allow knowledge and innovations to be freely available to the market. This not only would provide the benefits of a competitive market and afford public transparency and accountability, but it would also remove unnecessary burdens upon the taxpayer to fund bureaucratic systems of regulation and their enforcement. It would also remove the obstacles to the free flow and exchange of knowledge and innovation that is a condition for scientific creativity and equitable opportunities for international and social development. Providing that public universities and research institutions were democratically controlled at a municipal level, this would both decentralise and socialise scientific research and technological innovation, without requiring a centralised bureaucratic administration, and further liberalise the market. This would not prohibit private research, but it would not afford it any public protections, at public expense. Once private research enters the public realm, with its concomitant public effects, then it would become public property, freely available to all.

Commercial secrecy and the ability to suppress patents are not compatible with either democratic oversight or an open society. The success of scientific research and technological development are more likely in an open society wherein researchers can share knowledge and skills, leading to cross-fertilisation between scientific fields and receiving critical evaluation at all stages of the research, and, therefore, it is of benefit to scientists that their research should be made public as soon as possible, and, when research impacts upon the public then it is fairly obvious that, in a democracy, it should be under public scrutiny. Of course, scientists need a working environment that is free from interference and distractions, but the public should be involved in the choices of research funding, the localisation of research laboratories, and the implementation and development of research outside the laboratory. There are also benefits to scientists that arise from public involvement by increasing the stock of skills, knowledge, experience, imagination, creativity, and lateral thinking available to the researchers. It also shifts the parameters for critical evaluation away from those that had been taken for granted. An increased public awareness of the problems, limitations, and potentials of science would clearly be very helpful to both the public and researchers. This would be greatly enhanced by more public education about the sciences and scientists would have opportunities to effectively communicate their research to the public. Both scientists and their fellow citizens could benefit from more education in science studies in general, as well as reflecting and becoming aware of each other's concerns and needs. This promises to further develop a scientific culture, which would motivate children to study sciences and, thereby, increase the societal stock of highly motivated and educated scientists, and help overcome the cultural divide between science and humanities education. As C.P. Snow argued, it is essential that we get beyond the "two cultures" education system.[35] Even if children were not particularly interested in becoming a scientist, they would, as adults and citizens, be better placed to participate in deliberations and decisions about scientific education and research. Education and communication are crucial for the democratisation of science and creating the conditions for an open society. A close relation between communities, schools, colleges, and universities, would affirm public participation in the choices of studies and research that should be promoted and encouraged, as well as helping ordinary citizens improve their level of education and capacities to effectively communicate. School children should be involved, from an early age, in discussions about their visions for the future development of their communities. The public should fund the further education of college and school teachers through local university programmes studying the relations between science, technology, and democracy. Universities should research and teach the wider historical, sociological, and anthropological uses and adoptions of sciences and technologies in other communities, societies, and cultures; researching in detail the ways that sciences and social structures transform one another and researching public concerns and interests about the wider issues of education and communication. Moreover, through education, an increased public awareness of the history of the choices made during the ongoing construction of our society will increase the public awareness of the contingency of those choices and raise awareness of possible alternatives. This will not only increase the public confidence in its ability to choose, but will also help people anticipate the challenges, difficulties, and resistances that may well result from those choices.

Social pluralism not only leads to controversy regarding the operation and function of any new technology, but also provides tension, conflict, and the need for negotiation regarding the social perceptions and expectations about the social need and risks of any proposed innovation and its alternatives. However, this social pluralism is not an inefficient obstacle to technological innovation and social development, but is rather a fundamental precondition for the diversity and flexibility within society that is necessary for the possibility of open-ended decision-making that prevents the technological society from becoming totalitarian, inflexible, and, ultimately, blindly mechanistic in a complex and changing world. Heterogeneity is the condition for the intelligence, adaptability, and creativity of society providing that society embraces and respects that pluralism is a public good with practical value for societal development. The case for a democratic participation is strengthened by the argument that widespread social participation in decision-making increases the available stock of tacit knowledge, experience, and imagination upon which such decisions are to be made, thereby, enhancing society's capacity for dynamic, spontaneous, and localised re-organisation in response to unforeseen events, without the need for a centralised and bureaucratic hierarchy of decision-making and communications. Moreover, due to the pluralism inherent to a genuine democracy, the case for democratic participation is also strengthened by the argument that widespread social participation increases the social capacity for lateral thinking about ends as well as means. As contingent social products, every science and technology reflects social choices, negotiations, struggles, and compromises, within which the exploration of one set of possibilities is favoured instead of others. This is inevitable, but, what is at stake is the question about which sections of society are to participate in the struggles and negotiations to make the choices and compromises. As social circumstances change when new knowledge and innovations become available, local communities will need to re-evaluate current practices and expectations. It is impossible to provide a universal set of evaluative criteria that would be adequate for the circumstances and particular character of every community. Each and every community needs to be able to change and adapt its evaluative criteria in order to intelligently respond to the changing world within which it is situated. An awareness of local knowledge, character, and circumstances are essential if technical changes are to be incorporated and integrated into any community in a way that has the greatest likelihood of being sustainable and of benefit to the local citizens. Given that ordinary members of any community are better placed to evaluate their needs, circumstances, and community character, than any federal bureaucrats, professional politicians, or specialised scientists, then the process of deciding evaluative criteria is better left to local citizens. The bounds of technically rational decisions need to sensitively encompass local decisions and engage with a plurality of ends and means in order to optimise the sustainability, diversity, and flexibility of society in general. It is for this reason that it is of paramount importance that the technological society is democratised in order to maximise the diverse stock of tacit knowledge by liberating the creative potential of each community to critically evaluate local ends and the proposed means to achieve them. If we accept that pluralism is a social good (of practical value, as well as of moral worth) then we need to also accept that local, participatory, democratic processes by which heterogeneous elements are converged to solve our problems, in accordance with a principle of optimising the plurality of our society, should be the minimum requirement of any definition of the rational criteria for deciding which technologies to locally implement and develop, while it is essential that all citizens are able to satisfy basic human needs, such as access to food, water, shelter, education, healthcare, communications, and knowledge, if there is to be the possibility of creating the social conditions for a democratic society.

In a genuinely democratic and open society, the public also need access to publicly funded and locally controlled media and communications. The Internet and the World Wide Web (like PBS radio and television) have the potential to provide media and communications for an active citizenry, concerned with being well informed, as well as improving the capacity for public deliberation in democratic assemblies, and, as such, public media could constitute virtual assemblies, alongside neighbourhood, regional, national, and international assemblies. The Internet and the WWW offer the potential for widespread public participation in deliberating science and technology and show how information and network technologies can be democratically appropriated and transformed in ways that were not intended or anticipated by their designers. Both the Internet and the WWW were developed using taxpayers' funds, in government funded research institutions, rather than by the market. The Internet was developed from the ARPANET created by the US Department of Defence and the WWW was created and developed at CERN in Europe. Since the early 1990s, many social commentators and theorists have claimed that these information and network technologies have the inherent potential to democratise communication and the dissemination of knowledge by promoting the growth of virtual communities which will radically democratise society by providing grass-roots media for collective action based on free expression and association.[36] However, even with their democratic potential, we cannot rely on the Internet and WWW (and other forms of media) to provide grass-roots media, allowing free and democratic commun-

ication between citizens (including scientists), given their increased commercialisation through the corporate consolidation of network technologies.[37] Barney argued that a close inspection of current tendencies
in the development of the Internet, in terms of commerce, corporate
media consolidation, advertising, outsourcing, and surveillance, shows
that the Internet and WWW are actually intensifying the antidemocratic
and totalitarian aspects of industrial capitalism, mass consumerism,
and state control.[38] As he put it,

> Rather than any revolution... the proliferation of network techno-
> logy represents an acceleration of the logic and effects of capitalism
> in the practices and relationships of production, work, consump-
> tion, and exchange. It does so because of its peculiar properties
> – among them its ability to collapse control, information, and com-
> munications utilities into a single undifferentiated stream of bits
> – enable and encourage capitalization on the fertile environment for
> profiteering and accumulation created by transnational economic
> liberalization, privatization, and deregulation.[39]

Network technologies have become media for transaction rather than
interaction. The globalisation of network and communication technologies – consolidated in the hands of a few multinational corporations
– tends to reinforce the *status quo* rather than afford opportunities to
challenge it. Network technologies act as a structural resistance to democratic participation when they cease to be public property (funded through
taxpayers' contributions) and are privatised as the means of sustaining
and extending power relations and commercial ambitions. Hence, Barney
argued that

> Network technology enables the further entrenchment of those
> inequalities and control over the means of power that frustrate the
> equal ability of citizens to participate in the fundamental decisions
> that affect their everyday lives.[40]

Borgmann also considered network technologies to be little more than
an extension of the device paradigm, entrenching us further in the governing patterns of production and consumption.[41] In *Crossing the Postmodern Divide*, he developed his critique of the device paradigm further
and argued that network technologies not only channel us deeper into
patterns of consumption, but they reduce our capacity to be attentive, to
engage, and to communicate.[42] However, Borgmann has been criticised

for neglecting to examine how network technologies have the potential for positive reconstructions and enhancements of social being and personal experience, which cannot be reduced to simply the thrill or glamour of the use of devices, when they genuinely enhance the enjoyment of the present and make civic participation possible for disabled or isolated people.[43] While I agree that network technologies do have this potential, especially for disabled or isolated people, the criticisms made by Barney and Borgmann should be taken very seriously indeed. Whether the Internet and WWW are used to enhance education and communication or as a totalitarian means of propaganda (political or commercial), whether they are used to augment creativity or to transform the creative mind into a commodity, or whether they are used to liberate human beings or enslave them, depends entirely how we participate in the development of these technologies. Whether or not their democratic potential will be realised, or whether governments and corporations will control and dominate their development and usage, depends upon whether we are complicit consumers or critical citizens. As Feenberg and Barber noted, even though network technologies can be considered to have an inherent democratic potential, whether this potential will be realised depends on factors that transcend the technologies themselves.[44] The democratic development of network technologies will depend on the public ownership of these technologies alongside the political will to provide democratic oversight of their development. Steve Schneider has proposed that "federal assessment courts", comprised of elected scientists and public representatives, could provide democratic oversight for network technologies, as well as for biotechnologies, nuclear power, and other kinds of scientific research and technological development.[45] It would also be possible to further develop the public consultation committees, such as the "consensus conference" pioneered by the Danish government in the 1980s, or "voter juries", as a means of providing public participation and democratic oversight in formulating public policy for scientific research and technological innovation.[46] However, the practical value of democratic participation would only be recovered if such courts, conferences, or juries were able to maximise the diversity and plurality of perspectives on these major public issues. It is unlikely that any small group either elected or selected at random from the general population, would be able to achieve this. Public deliberations should involve the whole population. Given that participation would be voluntary, deliberation should be conducted through democratic assemblies of elected delegates directly accountable to neighbourhood or municipal assemblies, as well as through polyarchic organisations, and public deliberations on publicly owned and locally controlled media, such as PBS television, radio, and Internet based networks. Providing that they were decentralised, publicly owned, and locally controlled, cooperated, Barber's idea of Civic Communications Cooperatives (CCCs) would be an effective aid for participatory democracy. Citizens would be able to use CCCs to access the Internet and PBS to gather information and communicate independently of their elected delegates, as well as monitor the activities of their delegates.[47] This would mean that delegates would not be able to control information and manipulate their constituents, while public media would remain free of central control, in order to allow citizens access to diverse media, which constitutes the best safeguard against propaganda and other forms of media manipulation.

Democratic participation in science and technology

There is a commonplace assumption that lay citizens are simply not qualified to make judgements about the directions and content of science and technology. For example, Levitt and Gross asserted that

> Scientific decisions cannot be submitted to a plebiscite; the idea is absurd. Applied to science education, for example, letting people vote on what should be taught would give us countless schools in which "creation science" would replace evolutionary biology.[48]

There are three basic assumptions that underwrite this assertion. Firstly, Levitt and Gross assumed that large numbers of people would actually choose "creation science" over evolutionary biology. Do large numbers of people want "countless schools" that teach creationism rather than evolution theory? Where is the evidence to support this claim? It is not at all evident that the majority of people would make such a choice, even in religious societies and communities. Only a few American school boards have chosen "creation science" instead of evolutionary biology, or have insisted on caveats to be stamped on textbooks, but it has often been the parents who have been the main obstacle to teaching "creationism" in public schools by appealing to the authority of the courts on the basis of 1st Amendment of the US Constitution, which prohibits state sanction of any particular religious doctrine. It has been the democratic participation of parents, funding and making legal appeals to the courts, appealing to a political document in the name of preserving intellectual and religious freedom, rather than appealing to the scientific merits of *Origin of the Species* over *The Book of Genesis*, which has preserved the teaching of evolutionary biology in public schools. Secondly, Levitt and Gross have assumed that parents should not decide what children are taught in (a public) school. This assumes that the content of the curriculum should be a technical judgement, even though choices regarding the purpose of education involve value judgements, for which there is not any technical expertise. Thirdly, Lovitt and Gross assumed a pessimistic and simplistic representation of how "a plebiscite" would behave in practice. This imposes a very narrow conception of the form and structure of democratic participation. It is not the case that democratic participation necessarily involves forming "a plebiscite" to vote on the content of science education on every school curriculum, every scientific research proposal, or every technological project. Democratic participation involves the public participation of concerned citizens (including scientists, technicians, parents, teachers, students, and taxpayers) about particular research proposals, technological projects, or school curricula, but it does not require that citizens make their decisions within a consensus based system of democratic centralism. In practice, public participation involves a polyarchic complex of organisations, institutions, conferences, boards, and individuals engaged in public deliberations regarding education; the development of relations between science, technology, and public goods; shared visions of an ideal society; the moral obligations and limits of scientific research; the relations between religion and science; and questions about how scientific research should be directed, funded, conducted, and communicated. Democratic participation would involve taking steps to ensure that these deliberations between such organisations better reflect the value judgements of the vast majority, rather than those of a few powerful minorities or individuals. While a democratic consensus can be often achieved at the local level, within particular communities, it is at the societal level that the nature of democratic participation would form an irresolvably pluralistic and diverse social endeavour, involving competitive struggles and cooperative alliances, with over-lapping or conflicting concerns and interests, wherein the whole of society forms a polyarchic plebiscite, rather than some unitary "plebiscite" voting to impose "majority rule" upon the content of scientific education and research.

Democratic participation involves increasing public participation to include all citizens who wish to be included, at whatever level they are willing and able to participate, to help deliberate and decide matters of concern to them and their fellow citizens. Far from being absurd, it is necessary that the directions of scientific research and technological innovation are publicly subordinated to broader societal concerns and criteria about the public good(s) at stake. What is absurd is the assumption that specialised scientific and technical knowledge conveys expertise in understanding the public good, as well as providing the degree of foresight required to know how to realise it through the implementation and development of science and technology in a complex, open-ended and changing world. Of course this does not mean that some "plebiscite" should vote on what qualifies as scientific truth or methods and then impose these decisions on scientists. Indeed, it would be absurd, but it would also be undemocratic because it would need to suppress the dissenting voice of scientists in the deliberative and decision-making processes and compel scientists to implement and develop methods and theories that scientists knew to be unscientific. Democratic participation does not require science by majority decree, but it does require genuine opportunities for scientists and their fellow citizens to be able to publicly discover a broader understanding of science. Studies have shown that citizens can successfully acquire a level of technical and political expertise required to successfully challenge and criticise hegemonies of expertise by developing testimony for media campaigns and public hearings, which has proved beneficial for developing scientific research and the disseminating of scientific knowledge and new technologies, providing that citizens are given genuine opportunities to communicate their experiences and concerns to scientists.[49] By doing so, citizens not only challenge scientists to demonstrate the public good of any scientific proposal to the satisfaction of the public (who are funding scientific research either as taxpayers or consumers), but can lead to scientists questioning their own assumptions about science. Excluding the public as an *a priori* simply leads to badly implemented and developed science. Far from being an unwarranted interference in science, public participation can expose political and economic interference in the scientific process by demanding greater accountability and challenging assumptions. Consider the example of nuclear power. The public trust of scientists and professional politicians has been clearly eroded by the numerous cases of nuclear leaks and attempted cover ups, as well as by major accidents such as Three Mile Island and Chernobyl. Such cases show that there is a clear contradiction between the actuality of the performance of nuclear technology and the evaluations of risk that were made by scientists and used by politicians to claim that nuclear technology was a safe means to generate electricity. Public scrutiny can increase the potential for the rational development and implementation of scientific research and technological innovation, in the sense of producing better quality and more stable technical results, in the wider world outside the laboratory or beyond the drawing board, as well as better addressing, communicating, and understanding a broader range of public concerns.[50] Also consider genetic science and biotechnology. As Lori Andrews has argued, political and technocratic decisions have seriously undermined the broader understanding of the implications of genetic science and biotechnology by underfunding efforts to explore these connections and implications.[51] A very narrow conception of genetics and biotechnology is being imposed at the expense of a broad and rigorous understanding of these sciences and their implications. By broadening the critical criteria for evaluating the presuppositions and assumptions of science, its potentiality for objectivity can be recovered from its reduction and distortion to being the exercise of instrumental reason within a very narrow specification of criteria. If this potentiality for objectivity can be recovered through democratic participation then the objectivity of science, even as a critical ideal, is crucially dependent on the openness of the society within which science is situated, developed, and differentiated, and, consequently, the objectivity of science depends on the democratisation of wider society and the critical engagement with a democratic citizenry with the public understanding of science and technology. Democratic participation promises to positively improve science by subjecting it to an increased level of conjectural and critical deliberation, through public evaluation and debate about the nature and purpose of science.

We often accept that there are legitimate grounds for the public to place limits or conditions on the direction or limits of scientific research. It often taken to be legitimate for the public:

(1) to limit the allocation of limited public resources to scientific research. Physicists may well want large sums of taxpayers' money spent on building the next generation of particle accelerators, but it is reasonable for the public to prefer to spend the money elsewhere, e.g. on schools or hospitals;

(2) to direct more public resources to one direction of scientific research rather than another, even if those directions have equal scientific value e.g. oncology rather than astrophysics;

(3) to limit or prohibit methods of scientific research on moral grounds e.g. prohibiting experiments on human beings, such as genetic modification or cloning, or placing strict conditions on experiments on animals;

(4) to focus public resources on one particular area of science e.g. medicine;

(5) to regulate or control which institutions can conduct scientific research which has public risks, e.g. biotechnology or nuclear physics;

(6) to focus public debate in specific areas of scientific research, e.g. global warming or the dangers of smoking tobacco, even if other areas have equal scientific value.

However, it is widely taken to be illegitimate for the public to decide the outcome of scientific research. While the public can legitimately set goals, provide resources, and impose limits, it cannot legitimately decide what the conclusions of scientific research are on the basis of whether they are culturally or ideologically acceptable or not. Such a decision is widely taken to be overstepping the legitimate boundaries of democratic participation in scientific research. It is not the case that expertise needs to be insulated "against the vagaries of democracy", as Galston put it, but, rather, if we agree that scientific knowledge and technical expertise are public goods, then public self-restraint regarding the outcomes of scientific research is the best democratic means to realise the public good.[52] However, when expertise is specialised, in an open-ended, complex, and changing world, it cannot convey a total understanding of the consequences of any course of action. Nor does it convey any knowledge of the goodness of the goals to which technical expertise is applied to provide the means to realise them. Technical expertise must always be subordinate to democratic processes involved in deliberating and deciding the public good, which will include proposing directions and limits for scientific research and technological innovation. It is also the case that expertise is frequently contested, even in narrow areas of specialisation, and, therefore, the public acceptance of claims made by experts is not based on public judgement and/or knowledge but is a matter of public trust and consent based on credibility and public record. The democratic process overrides claims to expertise in the sense that any such claim needs to be subjected to public scrutiny and justified to the satisfaction of the public. The public trust of scientists and the objectivity of scientific research is jeopardised when scientists ignore and override local knowledge, experiences, and concerns. As Brian Wynne argued,

> The issues and problems of public understanding of science cannot... be divorced... from the epistemological issues of the social purposes of knowledge, and what counts as "sound knowledge" for different contexts. These in turn highlight questions about the institutions of

science – its forms of ownership, control, and practice. To preclude these issues from public debate is to undermine the possibilities of effective public uptake and culture of science.[53]

By imposing their own "scientific" theories, methods, and estimates (involving a host of assumptions made in the abstract) on the analysis of the impact of radiation from Chernobyl on the sheep of Cumbria, UK, and ignoring the local farmers' advice about how to conduct the research and take measurements, the reliability and objectivity of the scientists' conclusions were more science fiction than fact. The act of ignoring the local farmers was perceived by the farmers to be arrogant and this caused a high degree of friction and resentment which was translated into local resistance to the efforts of the research scientists. This further reduced the scope and quality of the scientific research. It is also the case that scientists tend to adopt the same cultural assumptions and prejudices as their fellow citizens, and, therefore, it is unreasonable to expect scientists to display a greater ability to challenge these assumptions and prejudices than expected of any citizen.[54] Even when scientists scrutinise the assumptions and prejudices underlying the work of their fellow scientists, as well as critically reflecting upon their own, we should not mistake sincerity for objectivity, at least not without radically reappraising our understanding of how the word "objectivity" is being used. It is highly problematical, if not arbitrary, to claim that scientists are objective about the prejudices and assumptions, which underwrite scientific methodologies and aspirations, on the basis of their sincerity combined with technical expertise in those methodologies. Even when scientists are intellectually virtuous, their work is only possible because of a history of the prior efforts and decisions of others, involving a series of assumptions and prejudices, as well as the criteria of success and goals imposed by those who fund scientific research. On the basis of their technical training and experience alone, scientists are no better placed to objectively understand the cultural background and history within which their work is situated than any other citizen is able to, and, if this is true, then it is unreasonable to consider a scientific training to be the prerequisite for deciding upon the goals for scientific development. This does not mean that the technical judgements of scientists about the possibilities of their own work should be ignored, but, it does not mean that such judgements are sufficient either. Hence, the societal development of science must permit the "lay citizenry" to challenge scientific consensus, by challenging scientists' right to lay claim to objective knowledge or truth; assumptions; methods; choices of directions of research, development, and implementation; and, their interpretations. These challenges can be made by publicly examining the historical development and differentiation of science; questioning the current scientific paradigm and exposing anomalies; calling on scientists to explain and justify their assumptions and methods; and also by exploring alternative kinds of knowledge alongside scientific knowledge.

If any technology is to be successful in solving problems in the wider world outside the laboratory or computer simulation, then it needs to be open to being constantly revised as it is implemented in context. Thus the best judges of the efficacy and limitations of any technology are those that use it and suffer its consequences. It is the involved public that is best suited to judge whether and how any technology works or fails. Thus, contrary to the current model of "representative democracy", the success of any technology is dependent upon its capacity to being dynamically flexible to pluralistic evaluation and experimentation in accordance with the experience and ideas of the people who actually use it. In order words, participatory democracy is essential to the development of the technological base of society. The participation of local citizens is essential to the successful implementation and development of technological projects. A failure to allow local participation in design and implementation suppresses the local knowledge and experience necessary for the success of technological projects.[55] The advantage to public participation in research and development is that the compatibility of any innovation with the wider public interests and concerns, as well as already existing social structures and practices, should become apparent much sooner when the public is involved from the onset, rather than only being involved during the implementation or contracting stages, when all the crucial technical decisions have already been made. Early public participation will reduce the total costs of research and development, given that modifications or even scrapping a project at the end stage are enormously costly, even more so if we include the public costs of social problems that have been caused by ignoring the incompatibility of implemented scientific or technical projects with public interests and concerns, but it also increases the chance that the implementation and development of any innovation will be stable, enduring, and predictable. Broadening the criteria of evaluation increases the chance of anticipating potential problems and complications, and, by including the public from the onset, there is a greater chance of developing a high level of trust between the public, politicians, bureaucrats, and scientists, which reduces the risk and cost of public resistance or opposition. Citizens' participation in the broadening of the criteria for the design process should be supported by technical experts rather than be eclipsed by them. While decisions about whether and how to implement any technology, say a hydroelectric dam, for example, should involve a host of expert testimonies from engineers, surveyors, ecologists, environmental scientists, geologists, etc., the successful integration of all those testimonies into a process that aims to converge upon a clear and satisfactory decision is not a technical problem. It should also involve the testimonies of sociologists, historians, and even moral philosophers, but, most importantly, such decisions must be made by the local people, who will be effected by the project should it go ahead, after hearing these expert testimonies, as well as their own views, and deliberating upon them to their satisfaction. As Benello pointed out,

> there is no single correct technical solution to social questions. Customs, norms, ethical and social considerations inject themselves at every point, and groups made up of those affected are the only valid interpreters of such norms and values.[56]

Once we acknowledge an absence of expertise in such matters, we can recognise the practical value of democratic participation for the process of deciding how to organise and relate diverse opinions and specialised testimonies in order to practically decide and plan a best course of action, in the absence of certainty regarding the conditions for human well-being and happiness. Such a decision requires a host of normative and evaluative judgements, which are beyond the remit of any technical experts, and are tantamount to guesses and intuitions.[57] While we may hope that they are guided by educated guesses and informed intuitions, we would be well advised to recognise that the consequences of our actions are not determined by our intentions, no matter how well meaning or carefully considered they may be, and the test of the soundness of any judgement remains *a posteriori*. Even though we may well try to include technical expertise as a way of reducing our reliance on good luck, in itself that does not provide guarantees, and we still may apply that expertise incorrectly and also may have chosen the wrong goals in the first instance. Democratic participation has practical value for the development of science and technology when it provides broader and contextual information for the purpose of understanding the complexities of a changing and open-ended world. The democratic values of reasoned persuasion, dissent, and criticism, alongside cooperation with others and tolerance for diversity and plurality, have practical value at a societal level for increasing the adaptability and sustainability of the same social conditions under which science can flourish. The same democratic values that can create the decentralised, pluralistic, and diverse conditions for the maximisation of participation in deliberation and decision-making at the local level, where those who are likely to experience the consequences of decisions are the ones who should be responsible for making them, also provide benefits of cooperation, communication, and experimentalism at a societal level. This promises to increase the level of public creativity, dissent, and criticism at play in decisions regarding the trajectories of scientific research and technological innovation, while also broadening the criteria for such public deliberations and decisions. This promises to provide beneficial conditions for scientific and technical activities to be developed in increasingly rigorous and stable trajectories as a result of public participation, scrutiny, and debate about what constitutes good science and technology, including its methods and techniques, as well as its goals and limits.

Modern technological enterprises involve a compartmentalised division of labour between scientists, engineers, managers, workers, technicians, politicians, public relations, etc., within which very few participants can claim to have a clear and comprehensive understanding of the whole enterprise. The claim that democratic participation in the implementation and development of technology is implausible because the public lacks sufficient technical expertise and literacy becomes increasingly unsustainable once we recognise that often there is not a single person involved in even modest technological enterprises that has a complete grasp of the whole enterprise from a technical point of view. This claim only serves those who wish to limit the decision-making process to governments and powerful corporations, even though professional politicians, bureaucrats, and shareholders are probably as equally technically inept and illiterate as any other member of the public. Thus the claim that the decision-making process should be limited to technical and economic criteria is one that is designed to suppress the awareness of the social contingency, value-ladeness, ideological motivations, and vested interests involved in these decisions and the public tendering of contracts. Ordinary citizens would (of course) be at a technical disadvantage with an economist or engineer when discussing the economic or technical merits and consequences of any technological practices and products, but when it comes to the discussion of social goals then we are all on a par. The value judgements of the economist or engineer are as equally subjective and value-laden as the value judgements of any other citizen and, therefore, reliance upon economists does not safeguard a society against arbitrary and partisan economic or technocratic policies. In fact, it makes them more likely. The best way to safeguard against subjectivity and vested interest is to encourage as many people as possible, from every walk of life, to participate in polyarchic and pluralistic decision-making processes of our local communities and to implement those decisions in the particular communities that made them. It is the capacity for ordinary citizens to "think outside the box" that recovers the practical value of public participation in the implementation and development of technological and economic policy. Creativity and innovation occurs from a heterogeneous background; pluralism and diversity enhances the social capacity for creativity and innovation. Local citizens are best placed to learn what they need to know, set the agenda, contest the criteria, and implement the decisions that they agree upon. Human motivation and freedom is best optimised when we embody the decisions we make in our everyday lives. Careful decision-making is more likely when one knows that one is going to directly suffer the consequences of poorly or hastily made decisions. Thus it is not only important, in a democratic society, that the preparatory debates and analyses are open to public participation, but it is essential that non-technical criteria have equal footing in the process of evaluating the desirability of any proposed innovation or policy change.

Maximal democratic participation involves genuine opportunities for public participation at all levels and stages of deciding the directions of scientific research and technological innovation. It denies the epistemological legitimacy of any partition of the public process of deliberation and decision-making into an "exclusive realm of experts" because technical expertise must be subordinated to a broader conception of the public good for which there is not any expertise. However, maximal democratic participation does not involve scientific and technological activities being dictated by public decrees in "the name of the people" or by majority consensus. It does not mean that the internal operations of any technology are simply a matter of popular consent, as if we could all vote that pizza ovens should produce gold from dough and thereby make ourselves all richer, but it does mean that questions of whether nuclear power stations should be operated on a principle of safety or profits first, or whether nuclear power stations should be built rather than wind and tidal based power stations, should be subjected to democratic deliberation and decision. Even if citizens agree that nuclear power is the best available technology for electricity generation then it should be a matter of democratic deliberation and decision where power stations should be built, by whom, how much the citizens are willing to pay, who will be employed at the stations, and who will manage them. If different regions or cities are able to try different options and ideas for electricity generation then this will provide the whole society with an empirical basis for comparing the consequences of different decisions, thus increasing the stock of available knowledge, and any mistakes will be localised. Where there is an expectation of risk outside the region or city, say neighbouring regions or cities have concerns regarding the construction of a nuclear power station and the possibility of a nuclear accident, then the boundaries of localisation must be expanded to include all concerned parties. Once technologies are shown to work well in practice, they become readily available options for similarly placed communities to implement and develop in accordance with their own social needs, particularities, and circumstances. If technologies are shown to be inadequate then, providing society has developed in accordance with technical pluralism, there will always be alternatives. Even when particular technologies and local circumstances do not permit a great deal of operational latitude and flexibility, there is still considerable scope for pluralism and imagination on questions of policy. Advocating democratic participation is not simply the affirmation of a moral principle, but is in fact an important way of making the technological infrastructure of regions and nations more sustainable, diverse, flexible, and robust because they are better integrated into the structures of social life by allowing the very people most familiar with local circumstances and needs to make the decisions about how they are going to develop their communities and improve their lives. It is a way of improving the technological base of society by optimising the decision-making base of society. It allows the technological structures of society to be ontologically comprised of parallel and complementary technological practices, alongside diverse and pluralistic evaluations of what the public good is, which, due to their decentralised basis, permits considerable participation, structural flexibility, and experimentation.

6
Towards a Rational Society

In *Towards a Rational Society*, Habermas stated his commitment to the need for the subordination of technical rationality to a broader understanding of rationality discovered through democratic participation and communication.[1] From the onset, he assumed that the problem of how to develop science and technology in society was to be solved by ensuring democratic institutions and structures mediate the rational development of scientific and technological progress. As a result of democratic mediation, the bounds of rationality will be progressively tested and developed, providing that scientists and technicians do not overstep the limits of their specialisation and people understand the cultural meaning and practical value of science and technology. In this way, science and technology can be developed freely as a means to discover the truth and improving technical systems, while also being subordinated to the democratic processes of deliberating shared cultural meanings and understandings in terms of lifeworld experiences. Habermas' account of scientific and technical progress was consciously developed to support his theory of societal development. It was for this reason that he argued that the premises of social theory must be independent of the results of particular sciences and he was concerned with the use of scientific concepts and representations, such as those taken from theories of natural evolution, to explain society.[2] Following on from his earlier work, Habermas' analysis of the historical conditions and structures of the trajectories of scientific thought and practice was developed in order to explain them in terms of the moral and practical development of society. He interpreted particular developments and differentiations of scientific theories and practices in terms of their wider moral and practical utilisation for societal development. The societal test for any scientific theory or practice was whether it provided technically and practically useful knowledge. It is on the basis of practical concerns, evaluated in relation to the historical conditions and social structures of development, which provides scientific theories and practices with moral content, in relation to their practical value for the advancement of society and the improvement of the human condition. The rationality of societal development is historically conditioned and contingent upon the development and differentiation of the human capacity to assess the meaning or consequences of any action. The possibility of rationality is distorted or diminished unless the evaluation of consequences (strategy) is subordinated to the evaluation of meanings (communication). Habermas argued that the rationality of human agency is possible and progressive, as the complexity of society increases, providing that technical systems of media and exchange are subordinate to the lifeworld of human experience, meanings, and relations. If this distinction is not maintained then the resulting systematisation of the lifeworld becomes governed by strategic instrumental imperatives, which reduces the goal of achieving meaningful mutual understanding to increasing instrumentality (quantified in terms of the increase of money and power). The social communicative acts of reasoned deliberation are replaced by a system of exchanges governed by "steering media" and the capacity for critical reflection on cultural meanings becomes greatly reduced and distorted. The rationality of the epistemological and moral learning process depends crucially on the potentiality for individuals to achieve communicative action as the basis for the social development and differentiation of cultural meanings in relation to the experience of the lifeworld. Hence, Habermas argued that we need a stronger public sphere to allow greater democratic participation in communicative action directed towards the agreement of shared cultural meanings.

Habermas attempted to develop a theory of *communicative competence*.[3] All speech is directed towards the achievement of agreement, even if it is rarely achieved, and it is only in relation to this agreement, when achieved, that the truth of statements or value of any norms can be rationally judged. He termed this normative dimension of speech as "the ideal speech situation" and considered its achievement to be the condition for the possibility of a rational communication act. According to Habermas, the structure of speech anticipates a form of life in which truth, freedom, and justice are possible. Even if these remain elusive, the possibility and meaning of speech depended on their possibility, which embodies normative standards in the structures of language and social action required to achieve agreement, without coercion or domination. On the other hand, ideology, in Habermas' terms, is a system of beliefs, prejudices, and norms that require coercion or domination to maintain their "legitimacy" and, consequently, the coercive conditions for maintaining this system suppress and distort the conditions for "the ideal speech situation". Communicative competence is a social development of recovering the possibility of "the ideal speech situation" by emancipating, through reflection and criticism, the structures of language from ideology and forming the basis for the exploration of developmental and evolutionary possibilities. Habermas presupposed the rationality of the societal gamble and considered scientific knowledge to provide the conditions for the technical mastery of society and the natural world, as well as organise human relations, expanding the sphere of "sensuous human activity" and *praxis* through work and communication, which allows the further development and social evolution of "forces of production" and "normative structures of interaction". For Habermas, all knowledge is historically developed, differentiated, and bounded by human interests. Scientific knowledge is the product of organised labour, conditioned by the practical understanding of the conditions of human existence and how to manipulate them. The historically developed and differentiated organisation of labour in relation to the material conditions of human existence determine the structure of objects of possible experience, as constructed through human activity, such as tool-use to manipulate and measure objects, and language-use to represent and communicate the understanding of those objects. Habermas argued that the "species being" of human beings provides an interest in the production and dissemination of the means to facilitate manipulation, calculation, and communication, in order to reproduce and develop the material and cognitive conditions of human existence, enhancing human freedom by emancipating the "species being" from material and cognitive limits, and making our own history. Work involves the orientation of technical control over objective processes, coordinated through strategic and purposive decisions regarding the efficient use of scientific knowledge in accordance with the calculated pursuit of human interests, coordinated and interlocked in a framework of strategies, norms, and intersubjectively recognised rules of procedure. Communication involves the orientation of mutual understanding, coordinated through intersubjectively recognised norms and rules required for reciprocity and the achievement of agreements. The sciences allow the formalisation and systematisation of work and communication, wherein reality is disclosed and acted upon through technical, practical, and emancipatory strategies for interpreting experience and developing social organisation. It is for this reason that Habermas rejected Marcuse's claim that science was inherently ideological.[4] Instead, he proposed that science and technology were inherent to the historical development of human beings as a species, but it is the domination of society in accordance with the class-interests of the economic elite, empowered through the technocratic systematisation of the application of instrumental reason, which has distorted the structure of human interests and threatened the possibility of rational communication and strategic-purposive action.

For Habermas, the public sphere was the democratic process of forming public opinion, ideally achieving "the ideal speech situation" by exercising freedom of assembly, association, and speech about matters of public concern and interest. Openness and criticality are essential for the justification of opinion through reasoned deliberation. This is a condition for public opinion to counter "mere opinion", which is the product of conformity to cultural prejudices, assumptions, and hearsay, as well as appeals to traditional dogmas and authorities. The possibility of "the ideal speech situation" has been undermined by the commercialisation of the public sphere, through the ongoing consolidation of private media and ideological privatisation of all aspects of public life, which has reduced it to a site of entertainment and propaganda. The public have been systematically excluded from participating in the formation of public opinion, except as consumers, and the potential for reasoned deliberation has been replaced with technical systems of information dissemination and analysis, publicity, public relations, advertising, and polling. This has weakened and distorted the social process of the development of rational communication within the public sphere. Science has become appropriated into a leading "force of production" within a global system of interdependent, large-scale industries, corporations, and the institutions of nation-states. This system has become the technical means to meet the "objective exigencies" produced by industrial capitalism, which has compelled the institutional arrangements of the State to be developed and refined into the political means to provide and maintain the conditions favourable to the global system and the economic elite. Governance has become reduced to a technocratic system of administration and regulation of public contracts, consumer rights, and the allocation of taxpayers' funds to purchase public services. The idea of political freedom has become reduced to a competitive mechanism for the selection (election) of administrative and regulatory personnel. The stabilisation of the conditions for the global system has become the only "legitimate" responsibility for politicians, bureaucrats, and technicians; the State has become the technocratic means to avoid risks and eliminate threats to the global system. By disassociating the criteria for justifying the organisation of society from a critical examination of normative regulation of work and communicative action, private media control has been effective in suppressing any social need for a distinction between the practical and the technical development of society. The suppression of public reasoned deliberation of alternative possibilities for societal organisation and development aids the ideological legitimation of the rationality of the pursuit of the particular interests of the economic elite, by representing these class-interests as universal human interests, which allows the technocratic domination of society (including scientific research and technological innovation) to be concealed under the mask of "technical rationality" or "objectivity". The technical imperative becomes an autonomous principle for the organisation of society in order to maximise control over society by the economic elite. Science becomes appropriated as the means of production of techniques to apply to specific purposes, as decided by a technical system for the optimisation of subjective preferences, while the decision regarding means is driven by the technological imperative to optimise efficiency, understood in terms of ongoing maximisation of productivity and minimisation of labour costs. Thus the global system has allowed the site of industrial "class conflict" to be successfully displaced from the so-called developed countries to exploited labour markets in the Third World, where industrial regulation and organised labour are either non-existent or uneven, and governmental interventions are based on overt acts of oppression and violence in order to create the conditions suitable for the global system.

Habermas assumed that science is potentially a value-free and open mode of communication in modern society, allowing scientists to share a commitment to the discovery and communication of the truth, but it has been distorted by the global system of media and exchanges in order to serve the class-interests of the economic elite. He assumed that the scientific world-view can be formulated at the level of communicating lifeworld experiences, discovering and using shared meanings, with practical value for human "species being", but has been constrained and distorted as it has entered into the "institutional core" wherein scientific results are used instrumentally in accordance with their technical value for the development and differentiation of the "forces of production" and the ideological justification of the global system through "the selective exploration of culturally available cognitive potentials".[5] However, once we have shown that science is not quite as value-free as Habermas assumed, we have shown that the boundary between the lifeworld and system are permeable in a modern society informed by scientifically developed and differentiated meanings and experiences.[6] This boundary has its own history and, as a result, the possibility of the identification of any objective conditions or structures for a rational communication act are also historically conditioned and contingent. Modern science is premised upon the representation of the world as a system and it is tested in terms of its technical value for the systematisation of the world.[7] In this respect, science is a systematisation of communicative action, an example of "steering media", and provides the lifeworld with specialised meanings and shared experiences, and Marcuse's claims regarding the ideological character of science have considerable merit. Once we take this into account, any conception of "progressive rationalisation" is itself at stake in the ongoing deliberative efforts, based on a series of negotiations and interpretations required to deal with the irresolvable ambiguity inherent to identifying "rational communication acts", given the irresolvable ambiguity about the consequences of any action or meaning of any statement. The whole process of understanding the nature of rationality is contested within a pluralistic and diverse society, with historically developed and differentiated epistemological and moral traditions, leading to genuine alternatives for the question of the idealisation and experimental realisation of the rational society. This provides a societal stock of genuine alternative trajectories of exploration and evaluation, each providing an alternative (but over-lapping) set of cultural meanings and shared experiences, available for reasoned deliberation in terms of their practical value for developing communities within a decentralised process of the exploration of the nature of rationality and practicality through the democratic participation in the local development of communities by the people who live in them. If we agree with Habermas that social evolution is an historical process of replacing certain structures with increasing rational ones, making societies "the bearers of evolution", then we must also recognise that the selection pressures are also historically and culturally contingent on how everyday life is organised through practical and social activities.[8] In any pluralistic, diverse, and heterogeneous society, communities become "the bearers of evolution". The moral and epistemological progress of each community is a lifeworld experiment in the realisation of shared visions of the good life, human well-being, and how to realise them through coordinated and cooperative strategic action. If any shared vision is achieved through the "ideal speech situation", the developmental logic of any community becomes an historically conditioned and contingent voluntary and localised experiment in the discovery of how to realise the ideal society. It is when society allows these experiments to continue, while providing opportunities for communication between communities, learning from each others' experiences and coming together to form alliances to achieve cooperative endeavours, that such an open society can be considered to be attempting to realise itself as the rational society.

Human interests and the ideal speech situation

Habermas identified three primary human interests:

(1) Controlling and manipulating objects and processes in the world in order to change the material conditions of human existence;
(2) Achieving mutual, intersubjective understanding through communication;
(3) Emancipation from coercive and suppressive forces and conditions that distort or suppress reasoned deliberation as the basis for action.

His conception of rationality was based on the fulfilment of these interests, allowing human beings to reason and make decisions in the light of the available knowledge, rules of discourse, and social needs. Domination, repression, and the ideological framing of action (including discourse) impose arbitrary limits and conditions upon the human capacity for rationality. Human beings must achieve self-knowledge, through critical reflection and by understanding the historical development and differentiation of the cultivation and ideological formation of theory and practice, to achieve emancipation from these arbitrary limits and conditions by revealing the structures of distortion, repression, and domination, which then can be (hopefully) dissolved, deconstructed, or removed. Ideals are the consequence of the human ability to critically reflect upon our own development, in the light of historical insights, and, on that basis, act with greater levels of conscious interest in our own emancipation. Through consciously connecting theory and practice, through self-reflection, the primary epistemological standard becomes one of satisfying the primary interests, which allows coercive, repressive, and distorting forces to be identified in terms of their influence on the formative process of being human. Hence it is through critical self-reflection on this primary epistemological standard that the practical and emancipatory value of reasoned deliberation can be consciously grasped and socially agreed. These three primary human interests are thereby deeply rooted in structures of human action, experience, and language, providing the conscious connection between theoretical understanding and practical activity. Habermas argued that hermeneutic (historically informed and critical) philosophy dissolves the barriers and distortions to reasoned deliberation and a conscious understanding of life, and also shows that work and communication are the primary activities in the human struggle for existence, meaning, and emancipation. Should science be informed by hermeneutic philosophy then it will not be limited to generating empirical-analytical knowledge of descriptive laws and theories, but will be able to recover meaning, identify distorting, repressive, or suppressive structures of domination, and connect the nomological and interpretive aspects of science with self-reflective emancipatory interest. If science was informed by hermeneutic philosophy then it would form the basis for the rational and practical development of human well-being, without degenerating into social engineering.

In this sense, hermeneutic science could be properly categorised as being emancipatory, over and above the technical use and development of knowledge, and (contra Popper) both Marxist theory and Freudian psychoanalysis could be properly categorised as scientific, whereas physics and chemistry would be the product of structures of distortion and domination that reduced explanatory science to those providing technical knowledge for the manipulation and control over the natural world (only satisfying the first primary interest). A hermeneutic science employs dialogue in order to gather the interpretations and meanings associated with any object of inquiry, exploring possible historical explanations of how those interpretations and meanings became associated with that object (satisfying the second primary interest). By formulating a general theory to explain these historical associations, it is possible to identify possible structural sources of distortion, domination, or repression that lead to ideological presuppositions and prejudices about the object of inquiry (satisfying the third primary interest). This general theory can be "tested" by critically examining whether it has the capacity to dissolve (as well as reveal) distorted communication, leading to a broader understanding of the whole process of associating specific interpretations and meanings with the object of inquiry. By scientifically transcending this process of "testing" through its emancipatory and explanatory dissolution of the limits to critical self-reflection, one can explicate the epistemological standards (or meta-theory) by which general theories are formulated, connecting human experiences with human history, in terms of a theory of social evolution. This theory is critically evaluated in terms of its emancipatory and practical value for critical self-reflection on the human capacity for the rational reconstruction and recovery of meaning as it "reproduces [human] life both through learning processes of socially organised labour and processes of mutual understanding in interactions mediated in ordinary language".[9] Human interests exemplify an historical continuity of a struggle for existence within a natural world, as it is known to human beings through the constituting activities involved in knowing that world, and scientific knowledge is rational only if it is based on self-reflective contemplation and is practically beneficial for human adaptive strategies within a natural world. This shows how the rationality of relations of production are historically conditioned and contingent when, in the light of further scientific knowledge, human beings can learn that previously rational material practices are ecologically damaging, threatening the organic or climatic conditions of human existence, and, therefore, their continuance would be irrational. Rationality should be understood as the outcome of a dialectical process of emancipatory and explanatory development and differentiation, which cannot be adequately understood in terms of ahistorical imperatives, including inborn biological instincts or universal epistemological or moral standards. However, it is not simply the case that scientific knowledge is pragmatically confirmed by its practical success (which could well be achieved by conformity to structures of distortion and domination), but knowledge is confirmed by its emancipatory and explanatory value, as truth, for its potential for the dissolution of structures of distortion and domination, removing barriers to further inquiry and critical self-reflection. Habermas' efforts to develop a meta-theory of knowledge as a transcendental, rather than pragmatic theory were directed towards recovering the possibility of the rational reconstruction of the human potential for an enlightened transcendence and liberation from the historically developed and differentiated structural constraints and limitations on social being and its evolution. In this respect, Habermas rejected the assumption that self-reflection is a primarily subjective activity because he considered rational reconstruction, tested in relation to its enhancement of communicative competence, to be a basic aspect of self-reflection. This test is a social process of revealing and examining the rules of discourse, which makes ordinary discourse and specialised discourses possible, and removing obstacles to successful communication acts. It is through ongoing critical self-reflection upon the conditions for communicative competence that allows the dissolution of communicative distortions and provides the social conditions for the rational reconstruction of the possibility of "the ideal speech situation".

For Habermas, the possibility of "the ideal speech situation" was based on rational criticism and the identification of social procedures by which rationally motivated agreement could be attained. He argued that such procedures can be located in the structures and rules of discourse, wherein the presuppositions and procedures of discourse ground the establishment of truth and morals, providing that they satisfy "reciprocal behavioural expectations", which have been afforded normative status in terms of a common interest between all involved participants, arrived at through unconstrained reasoned deliberation. The rationality of discourse is dependent upon the possibility of arriving at common interests through unconstrained reasoned deliberation and, therefore, communicative competence is evaluated in relation to the proximity of the discursive circumstances to "the ideal speech situation". It is only on this basis that participants abide by the same rules of discourse, share meanings and interpretations, and achieve mutual understanding through intersubjective agreements. Of course, this raises the question of whether communicative competence is possible, beyond the case of concrete instructions with definitive outcomes (such as "please pass the salt") or in games with well defined rules (such as chess). Habermas has not provided a satisfactory answer to this question when applied to complex discourse involving abstractions, metaphors, and implicit meanings. Without an answer to this question, it remains irresolvably ambiguous whether human beings can fully identify and abide by the same rules of discourse and fulfil the conditions for communicative competence. If this is the case, we could not know whether the "ideal speech situation" had been achieved, or even approximated, we could not know whether reasoned deliberation was unconstrained, and we could not know whether our criticisms or agreements were rationally motivated or not. Arguably, even though Habermas' metatheory has identified the conditions for "the ideal speech situation", providing a standard against which we can identify distorted communication, he has not provided any universally acceptable criteria for knowing when these conditions have been met and "the ideal speech situation" has been achieved. In this respect, Habermas has only provided us with a principle of falsification, rather than a principle of verification. Indeed, it is plausible that the structures of language presuppose that we can relate statements to experience, thereby determining their truth-status, but it does not follow from this reasonableness of this anticipation that it will be satisfied, except in the case of simple, concrete instructions, wherein the communicative success of "please pass the salt" during dinner can be established when the salt is passed.

Perhaps it is simply the case that, for more complicated discourse, even if we expect or need to be able to correctly decide whether any communicative act is successful, we are unable to discover any universally agreed standards upon which to universally agree on the correctness of such a decision. It may well be the case that truth-claims must be made through references to "objective reality", but it does not follow from the ability to make such references to an abstraction that one has successfully verified any claim at all, except for a few simple, concrete cases.

How can we proceed? – In a heterogeneous society, diversity and pluralism create a dynamic tension wherein "rationality", "reciprocal behavioural expectations", and epistemological and moral standards are contestable. The irresolvable differences within such a society create criticism and dissent, which, if harnessed through democratic participation in deciding what constitutes "reasoned deliberation" and how to agree upon "the rules of discourse", provides the conditions for critically and rationally constituting the possibility of "the ideal speech situation" by perpetually challenging whether its conditions are adequately understood and have been met. It is the dynamic tension itself, created through pluralism and diversity, which provides the conditions for preserving the possibility of "the ideal speech situation". By publicly questioning and contesting, through democratic participation in assemblies and publicly accessible media, the meaning of ideas such as "freedom of speech", "freedom of association", and "freedom of assembly", they remain perpetually fresh and vital aspects of how the *demos* comes to understand itself as the *demos*. The public demand for maximal inclusion in a heterogeneous society provides the dynamic momentum required to overcome constraint, distortion, and domination by challenging the rules of discourse and societal complacency, as an ongoing process of increasing inclusion. Any social limits to inclusion are open to criticism by those who consider themselves to be excluded. Participatory democracy preserves the possibility of the rational society by maximising democratic participation by publicly questioning its openness as a civic society. The idealisation of the Constitution as the means for a realisation of positive political freedoms (of expression, association, and assembly) in relation to negative political freedoms (from oppression, suppression, distortion, and coercion) can only hope to be rationally grounded through maximally inclusive public deliberation and decision about the meaning of freedoms, rights, duties, and the Constitution itself. The dynamic tension in a heterogeneous society can create the possibility of an open public sphere wherein positive freedom can emerge, over and above its expression as will to power, within the boundaries of negative freedom, as defined though political action to maximise inclusion. Dissent and criticism are recognised as essential aspects of public, reasoned deliberation about the adequacy of current levels of societal inclusion. In this respect, participatory democracy preserves the conditions of its own possibility by embracing cooperative and adversarial forms of political action within democratic assemblies, directed to understanding the nature of "democracy" itself. Political freedom and democracy are the outcomes of political activity, rather than its condition, and it is only on the basis of maximally inclusive democratic political activity that we can hope to identify and realise "the ideal speech situation", and, thereby discover and satisfy human interests, over and above particular class-interests or the assertion of private preferences. It is essential that "democracy" and "freedom" are contested terms, in theory and practice; a shared, public understanding of these terms can only be an outcome of shared, public deliberation and political activity, rather than a condition for them. A common sense of purpose to democratic participation is something that must emerge from public participation in assemblies to deliberate and discover this common sense of purpose. Hence, as Arendt argued, communication is the condition for political freedom, within a public sphere created as a medium for speech between equals. In this sense, political freedom is premised upon the expression of social being within a shared language, through which choices become possible and we are able to define our responsibilities (duties) and respect for others as the social basis for the reasonable expectation of being capable of deliberating and deciding with others what the best course of action could be, without resorting to violence to impose our will on others. Pluralism, as the opposite of totalitarianism, provides the necessary conditions for political speech to discover "the elementary grammar of political action and its more complicated syntax".[10] This syntax is required for structured political power to emerge from "action in concert", as positive and non-violent participation, without resorting to threats or domination, premised on the making and keeping of promises to cooperate with others, forming a contract, compact, or covenant. Whereas pluralism necessarily finds its concrete expression of difference and dissent in a democratic council or assembly, totalitarianism necessarily resorts to the labour or concentration camp to suppress all difference or dissent. In this respect, "the ideal speech situation" does not simply refer to the political conditions for satisfying human interests, but, rather, involves a deeper reflection upon and understanding of the possibility of political action, free from distortion, coercion, and repression. It involves reflecting upon and understanding the conditions for and nature of distortion, coercion, and repression, and, to this end, Habermas' characterisation of critical, hermeneutic science offers a valuable philosophical heuristic for political thought and social theory.

Epistemological and moral standards remain irresolvably contestable within a pluralistic and maximally inclusive democratic society. The reasoned deliberation of such standards would lead to local experiments, conducted to the satisfaction of the individual consciences of the particular citizens concerned with the outcome and the practical value of such standards for living a good life. Political freedom becomes the best practical means to respond to irresolvable uncertainty and ambiguity in an open-ended, complex, and changing world. As Galston put it,

> Freedom operates in a zone of partial but not complete regularity, a discursive arena in which some reasons are better than others but none is clearly dominant over all the rest in every situation. If ethics and politics are part of this zone, as they evidently are, then their substance will reflect this ceaseless interplay of strong but not compelling reasons for grappling with the variability of practical circumstances.[11]

The chances of discovering and realising how to live a good life for the vast majority of people are maximised by democratic participation, in the absence of universal agreement about epistemological or moral standards, whereby each citizen is able to participate in the reasoned deliberation of any proposal or matter of concern to the satisfaction of their own individual conscience. The practical value of democratic participation can only be realised by applying the principles of free-association, localisation, and maximal inclusion within a polyarchic society that has already developed its communication technologies to a level where any citizen could contact and communicate with any other through decentralised media. Once this practical value has been realised then the effectiveness of decentralised democratic assemblies for the whole of society would be quantitatively and qualitatively enhanced by the local sovereign power of local citizens over the use and development of their skills, experience, knowledge, imagination, and resources. The transition from a liberal system of representation to a participatory democracy would be achieved when the vast majority of people not only aspire to hold sovereign power, but also expect it as the minimum condition for their voluntary participation in the workplace and their communities. Once this expectation has been turned into practice through cooperative endeavours then in turn it is turned into a practical way of life and the means for people to change the conditions of their existence. After its practical value has been realised in the organisation of the workplace and communities, the potential for democratic participation to be of aesthetic and ethical value as a pleasure and an intrinsic good can emerge and be realised through its practical value. Participatory democracy would provide the conditions for "living within the truth", as Vaclev Havel expressed it.[12] Havel argued that there were many parallels between the Western countries and the so-called communist regimes of the Soviet Union and the Eastern Bloc. The citizens of Western countries, while able to enjoy a level of consumption and security unknown to the people of the Soviet Union and the Eastern Bloc, are becoming victims to exactly the same kind of "automatism" as the victims of post-Stalinist communist oppression. Unless citizens of Western countries are prepared to change their patterns of consumption and value their freedom, moral integrity, and authenticity, over their desire for material comforts, then Western countries would inevitably tend towards a degree of totalitarianism on a par with the post-Stalinist communist regimes (such as post-Stalinist communist Czechoslovakia). Since Western countries have developed more sophisticated psychological and sociological techniques of public manipulation and control than the Stalinist terrorist techniques of arrest, imprisonment, torture, and murder of dissenters and critics, they have tended towards a more sophisticated kind of totalitarianism, which Havel termed as "post-totalitarian".[13] The post-totalitarian regime utilises subtle and all pervasive deceit, manipulation, and media control, alongside developing an implicitly coercive legal framework of state powers, to create a mass society based on conformity, uniformity, and discipline, wherein complicity is rewarded with improved levels of security and consumption, while resistance is punished by exclusion and suppression. Such a regime undermines the plurality and diversity necessary for the development of a free, open, and democratic society. A post-totalitarian society erodes any sense of civic or social responsibility among its citizens, apart from a sense of "patriotic" obedience towards "the Nation", while the concept of "freedom" is reduced to a sense of the personal ability to gain improved standards of security and consumption by conforming to the demands of the system of media and exchange (often labelled as "the Market"). Citizens in a "mass society" feel little personal commitment to any sense of authenticity, moral integrity (apart from their obedience to the given social norms and conventions), or freedom, especially if such a commitment might involve personal risk or discomfort. This absence of commitment is itself the product of education and media propaganda, combined with economic and political efforts to undermine and suppress all genuine alternatives to the system of media and exchange. It operates by creating the belief that "freedom" is identical with economic and political stability and security, and, therefore, any act that even potentially threatens the system can be represented as being against "freedom". This belief is necessary for the majority to consider its subordination to the interests to the economic elite to be both a private and public good, while it is in the interest of the economic elite to preserve this belief at all costs. In this respect, deceit becomes the fundamental basis for societal cohesion. However, as Havel argued, "living within a lie" does not necessarily involve believing that lie, but it requires behaving as if one does, turning a blind eye to that lie, and remaining silent about it when working with others who claim to believe it. It involves, as some level, being complicit with that lie within one's way of life. Hence the post-totalitarian society does not require that we actually believe the lies upon which it ideologically justifies itself in education and media, but it does require that we conform to living as if we believed those lies. Hence, political cynicism and apathy are acceptable within the post-totalitarian society, but dissent and civil disobedience must be marginalised, discredited, and suppressed.

The possibility of constructing a rational society requires direct, sustained, and widespread public participation in the deliberative and decision-making processes involved in critically evaluating the criteria for establishing the nature of rationality and the agenda for the development and implementation of scientific research and technological innovation. This involves a public understanding of how the structures of social power have been reproduced and intensified through structures of technological power embodied in societal infrastructure, policies, and strategies. It is essential for the democratisation of science and technology that the public develops strategies of resistance to the structuration of power within the technological infrastructure of society by appropriating those technologies and transforming them in accordance with alternative visions of society as an egalitarian, sustainable, and rational society. This involves the development of pluralistic and diverse visions of society and human well-being, creating genuine alternatives, as well as maximising opportunities for egalitarian public participation in the choice of which technologies to develop and implement in order to realise those visions. In this way, the construction of the rational society should be bound up with democratic participation in developing the conditions for community and individual well-being through the democratic structuration of the technological infrastructure and techno-scientific endeavours within society. Some kinds of technology are incompatible with democracy if they undermine constitutional principles or impose hierarchical structures upon societal development. For example, consider the example of the governmental use of data gathering and surveillance technologies on its own citizens in the United States of America. Any governmental use of data mining or trawling technologies to collect vast amounts of personal information and communications in order to search for and trace patterns, connections, or specific data from the general population, would be incompatible with the Constitution of the United States of America. These technologies are incompatible with the 4th Amendment, which reads:

> The right of the people to be secure in their persons, houses, papers, and effects, against unreasonable searches and seizures, shall not be violated, and no warrants shall issue, but upon probable cause, supported by oath or affirmation, and particularly describing the place to be searched, and the persons of things to be seized.[14]

Once "unreasonable searches and seizures" are defined as being those that occur "without a warrant", then the conditions of "upon probable cause" and "particularly describing" would make the use of such information gathering technologies, such as the notorious NSA Total Information Awareness program (TIA), unconstitutional because this type of technology, or any information gathering technologies that use a combination of database searches and pattern recognition software to compile a large database of information on the general population, functions by searching large amounts of data or communications, anonymously collected from large numbers of people, without regard for "probable cause" or "particularly describing".[15] Once a threshold level of information gathering is reached (for example the level of information gathering involved in effectively scanning all international telephone calls or emails to or from the USA for specific key words, numbers, or addresses; mapping out national communications networks by analysing all national telephone records for data patterns; or monitoring all bank transactions to and from the USA), then it becomes impossible to restrict such programs in accordance with any notion of "probable cause" or "particularly describing" and, hence, it becomes impossible to maintain effective judicial oversight and uphold the 4th Amendment.[16] This kind of technology also undermine the 1st Amendment rights to freedom of speech and association, in so far as it undermines any public expectation of privacy from governmental surveillance, and, thereby, undermines the possibility of "the ideal speech situation" and, arguably, would be incompatible with any kind of democracy. Putting aside all the possible abuses of executive power that such covert surveillance technologies make possible (such as spying on domestic political opposition or protest groups, eavesdropping on privileged communications between lawyers and their clients, identifying the confidential sources used by journalists critical of government policy, or even providing access to private commercial information for corrupt purposes), data mining and gathering technologies are inherently damaging to the democratic character of society because, even in the event that they are used only for the purpose of providing intelligence allowing the military and law enforcement to better protect the general public, they undermine the constitutional basis of its political and legal system by removing public oversight on executive power, and, thereby, the law becomes subject to the discretion of agents of the state. Under such circumstances, this undermines the independence of the judiciary as an important "check" upon executive power, and it also opens the possibility that institutional systems of law enforcement and espionage become the covert instruments of the incumbent administration. This places an inordinate amount of power in the hands of the political elite and undermines the possibility of successfully exposing corruption or overthrowing despotism.

Does this mean that any or all covert surveillance technologies are inherently undemocratic? What if the citizens of the USA, via their elected representatives or in a referendum, collectively agreed that their Constitution is no longer relevant to the conditions of the modern world and they would prefer the nation's police and military to have access to such technologies to better protect the public? Would this mean that they would cease to be a democratic society? What if the majority of citizens agreed to the provision that at any time thereafter, when circumstances were favourable, they could call for a referendum to suspend the use of these technologies and resume the constitutional basis for their legal system? The legitimacy of any such majority decision presupposes a unitary form of democracy based in consensus decisions, rather than a participatory democracy based on maximal inclusion and enrolment. Such a unitary democracy would be clearly sowing the seeds of its own destruction by building the foundations for its transformation into a totalitarian and oppressive regime, but would the decision to do so be a democratic one, if the vast majority agreed to it? Even if we considered this majority decision to be a democratic one, it is evident is that the character of that democracy would have profoundly changed. It would cease having the form of a Constitutional Republic and would tend towards state control thinly legitimated through democratic centralism. However, if there is the public expectation that any particular technologies would have an irreversible public effect on the structure of the whole of society, by changing the institutional organisation and power relations of that society, then such a technology would then irreversibly and dramatically limit the possibilities for the democratic participation of future generations. This may well lead to an irreversible rift between the structure and form of society to such an extent that the form is transformed into propaganda, preserving the illusion that the society remains true to its democratic ideals, while it suppresses these ideals and becomes structurally totalitarian. If this is the case, then, even if it were unanimously agreed by everyone to allow data gathering and mining technologies, such as TAI, to be used by military intelligence and law enforcement to protect the public, then, given that the use of such a technology would clearly be a secretive project (otherwise it would not work), it creates an inequality of power, centralising power and oversight in the hands of only a proportion of the population (those that have access to this technology), which means that, when the populace made their decision to support this technology, it placed an unreasonable expectation of moral commitment to the public good, on the part of those that have access to this technology. It would also assume that this empowered minority would agreed to relinquish their structural power and position if they are democratically called upon to do so in the future, and, consequently, by neglecting to include the consent of future generations of citizens, even though they would be effected by such a decision, such a majority decision would irreversibly exclude future generations from democratic participation in societal development. By making irreversible and uncontrollable changes to the structure of the totality of society, the citizenry would contravene the principles of free-association, localisation, and maximal inclusion. As a result, given that the current populace would be excluding the future populace from democratic participation, risking the imposition of a totalitarian police-state upon the future populace, then the decision to use such a technology would be an undemocratic decision, even if it was unanimous in the present. Such a decision would be an abdication (or a surrender) of democratic responsibility to preserve the form and structure of democracy. It is a fundamental condition for democracy that any public agreement is one that preserves public oversight and the capacity for democratic participation in dealing with the consequences of that agreement. Anything else is simply an abdication of democracy.

Democratic participation and socratic citizenship

Citizenship is a contested term.[17] As Ruth Lister has argued, the considerable body of academic and political literature exploring the concept of citizenships has tended to overlook gender inequalities and differences.[18] There are also vital cultural and local differences that also must be considered too.[19] Citizenship is substantively more than a set of constitutionally defined rights and duties, but is intimately and inextricably bounded by social identity within a civic society, which is itself the product of historically contingent struggles and alliances. This concept cannot be considered as politically neutral, especially when its uses presuppose societal norms and visions of an ideal society, which are not only historically and culturally mediated, but also important for constructing social identity, including nationality, family relationships, and the division of labour. The ideals and norms of citizenship are bound-together with the substantive meaning of community membership, which involves the ongoing development of civic norms and virtues, often understood in terms of the relation between civic society and the State, as well as the expectations that citizens impose upon their neighbours. Lister argued that we need to pay close attention to the uses of the concept of citizenship as a means to include some people and exclude others, within a discourse of social identity, differences, obligations, and resistance. The boundaries of inclusion and exclusion, drawn up to define the preconditions for citizenship, are an important site of social struggles and alliances. As Lister put it,

> While not denying the ways in which legal definitions of citizenship and citizenship practices can exclude (often block) outsiders and more generally as a disciplinary force, as an ideal it can also provide a potent weapon in the hands of disadvantaged and oppressed groups of insiders.[20]

The concept of citizenship is, in many respects, bounded by expectations and norms of human agency and how these relate to the rights and duties of civic participation. This is particularly apparent when citizenship is a condition for access to political and public welfare institutions;

involving discourses and cultural representations of social inequalities.[21] Hence Lister argued that,

> To act as a citizen requires first a sense of agency, the belief that one can act; acting as a citizen, especially collectively, in turn fosters that sense of agency. Thus agency is not simply about the capacity to choose and act but it is also about a conscious capacity, which is important to the individual's self-identity... The development of a conscious sense of agency, at both the personal and political level, is crucial to women's sense of themselves as full and active citizens on their own and in alliance with others.[22]

In a participatory democracy it is the conception of citizenship which is at stake – to be discovered through critical deliberation in assemblies of participants who hold each others' contributions to be equally worthy of consideration, publicly seek understanding, value intellectual virtues, and are concerned with discovering the best course of action through political speech. Once we understand that the practical value of democratic participation is that it facilitates the local exploration of particular epistemological and moral standards, in the absence of universally agreed epistemological and moral standards, the transition from the practical to the personal value of democratic participation becomes the existential basis for the simultaneous realisation of social being and radical individuality in every political speech and action. Participatory democracy promises to liberate the potential to recover humanity and individuality from conformity to "mass society" by recovering the practical and existential preconditions for human freedom to be realised through political speech as a social activity directed towards the conception and the realisation of "the ideal speech situation". A failure to do so will inevitably result in the totalitarian reduction of humanity to the performance of individual reproductions of "mass man" as a functionary within economic systems of production, exchange, and consumption that constitute the "mass society". The public, reasoned deliberation of how rationality is itself at stake in a participatory process of criticism and discovery is itself a condition for rationality. The "mass society" suppresses this process by distorting it into a utilitarian calculation of the sum total of individual consumer preferences – as the only legitimate expression of freedom – and, thereby, distorts societal development into a meaningless and irrational reproduction and extension of the "mass society". This tends to reduce the possibilities of societal development to the requirements of the systems of production, exchange, and consumption. However, as Arendt argued in *The Origins of Totalitarianism*, the public realm and the possibility of political freedom are suppressed and dissolved as soon as one perspective takes upon itself a monopoly over truth and the right to speak. The fundamental importance of the public realm is not that it provides a procedure for achieving decisions through consensus, but that it provides pluralistic and diverse perspectives, which comprises the condition for each and every one of us to think and judge matters for ourselves. Thinking and judging for oneself is the opposite of conformity and provides the basis for genuine civic responsibility between equals on the basis of the conscience and integrity of each and every citizen. In this sense, the public realm preserves the sanctity of the private realm by providing the distance and differences necessary for independent judgement regarding existential truths and moral commitments. It provides the conditions for individuality and intellectual conscience to lead to shared existential truths and moral commitments to emerge from political speech, without coercion, authority, or conformity. It is on this basis that participatory democracy promises to recover the Socratic sense of "citizens among citizens" in our shared, public pursuit of freedom, truth, justice, and happiness, especially when we do not agree what these ideas mean and how to apply them to our lives.

We need to reawaken the Socratic importance of the philosophically examined life and critically evaluate the meaning of how we understand and participate in the modern world.[23] If we are to authentically and meaningfully engage with our lives and reality, as thinking and critical beings, we also need to examine those aspects of the modern world that are actively and tacitly marshalled against, resist, and even suppress the very idea that we need to question and examine our lives and the way that we understand reality. We need to examine the way that we are situated from our birth as social beings and how society provides us with institutional and technological frameworks of mechanisms and resources, which transform us into an agent capable of participating in society. By taking the time to consider matters honestly, openly considering alternative views, taking responsibility for our choices, questioning our assumptions and presuppositions, criticising our prejudices and deeply held beliefs, and changing as a result of this examination, then – in the deepest philosophical sense – we are open to being members of a genuine democracy. In a participatory democracy every citizen would be an administrator, legislator, and magistrate – the Constitution would be an articulation of the "General Will" of the vast majority, as a social convergence of interpretations of the liberal tradition based on the outcome of critical deliberations about the nature of democracy, law, liberty, rights, justice, etc. Democratic participation would be best served by every citizen engaging in Socratic examination of his or her fellows – termed as "dissolvent rationality" by Dana Villa – in order to recover the practical value of dissent and criticism for challenging societal conventions and consensus regarding the nature of citizenship and civic virtues, which promises to achieve better articulations and understandings of them through critical reasoning and deliberation.[24] Conformity and complacency among the citizenry undermine the possibility of genuine democracy, especially in the modern era of mass media, electronic surveillance, and global communications, because they precondition the citizenry to propaganda and allow the suppression of dissenting and critical voices to be represented as preserving the norms and values of democracy. Without dissolvent rationality, any form of democracy in a modern society will inevitably result in a totalitarian mass society. By collectively giving ourselves a public voice, through cooperative democratic participation, testing communicative action through dissolvent rationality, citizens can subordinate technical systematisation to a broader and deeper understanding of public goods and concerns. Communicative competence, in this context, depends on citizens learning the cultural meanings of technical discourse and how this can be used for broadening the understanding of societal means. In this way, public participation actually contributes to increasing the chances of the pluralistic and sustainable implementation and development of scientific research and technological innovation. Children need to be invited to participate in decision-making and taking responsibility from the earliest age possible. Of course, this needs to be sensitively and carefully guided and nurtured by parents, teachers, and other community members. Education is essential for the development of a participatory democracy. The purpose of this education should be to teach children the civic virtues and social skills that they need to effectively, democratically participate, as well as the technical skills needed for societal development, alongside the intellectual skills needed to critically reflect and deliberate upon cultural and historical studies of their traditions and circumstances. It is essential that education develops children's capacity for critical reflection and deliberation as a fundamental aspect of the development of a well rounded character capable of challenging preconceptions and realising their potential as human beings in a complex, changing, and open-ended world. We also need to incorporate the questions about ecological sustainability into the public debate and understanding of cycles of production, consumption, and waste – linking the technical, political, and ecological within an adaptive model of bounded rationality.[25] Of course democratic participation in scientific research and technological development would slow down the process considerably, but coordinated and careful processes of public deliberation and decision-making, in accordance with community and workplace needs and capacities, would increase the chances of developing practical, coherent, and sustainable relations of production and consumption, as well as providing the potential for the rational inquiry into the desirability and possible public effects of any proposed project. This provides the possibility of scientific research and technological development benefiting the vast majority of people, while respecting local traditions and being ecologically sustainable. The commonplace prejudice against the demands of democratic deliberation is often stated along the lines of "democracy is all well and good, but there comes a point when discussion has to stop and we have to act." Once we consider the practical value of democracy, this prejudice is quite absurd because it assumes that discussion and action are independent, or at most weakly related. I term this prejudice "premature ratification". However, if we assume that the purpose of discussion is to decide upon the best course of action, the end of discussion should coincide with agreement about what course of action to take, and, consequently, only circumstances should impose time limits.

It has been argued that democratic participation requires the existence of a rational and committed citizenry capable of defending their own rights against the corruption and tyranny of powerful individuals and groups, as well as taking responsibility for the development of society. My position is that this argument is only correct in part. Only the commitment of the majority of people to learn how to cooperatively govern their own affairs is necessary for the existence of democratic participation, but rationality is a contested idea. There is an absence of any universal agreement about what is involved in being rational, whether it is possible, or even if it is desirable. In the absence of any such universal agreement, it seems to me that rationality should remain at stake, preserved by the dynamic tension and dissent inherent to public deliberation and decision in a heterogeneous society, rather than an *a priori* condition for democratic participation. It may well be the case that such a process is an unending exploration of rationality that is itself perpetually open to experimental development and critical examination. If this is the case then rationality can be an aspiration, an ideal, which has meaning for our critical capacity to resist complacency and conformity, even if we can never realise that ideal in our practices and discourse. Even if it is possible to discover how to be rational, the possibility of learning how to be a rational citizen is a potential mani-festation of social being which must be discovered through democratic participation, as an evolving process, rather than a condition for it. Furthermore, the rights of the citizenry should also be the outcome of public deliberation and agreement, as part of the social process of discov-ering the meaning of citizenry, rather than a set of *a priori* constitutional principles, which are not subject to revision or refinement, defined by some intellectual minority, acting as a "revolutionary vanguard" or "founding fathers". If the Constitution is to remain meaningful as a General Authority then it must be a social contract between the vast majority arrived at through public deliberation, as an expression of how the *demos* understands and defines itself, rather than a condition for that understanding. The meaning of democratic participation is to be dialect-ically discovered through public processes of deliberation and agreement about how to cooperate, by experimenting and negotiating with each other about how to improve our communities and workplaces in a way that increases our chances of discovering how to live a good life. In other words, it is only through democratic participation that human beings can learn how to collectively discover, articulate, affirm, and defend their shared vision of a good life, take responsibility for the development of society, define their rights and the meaning of citizenship, and awaken the potential for rationality. The greatest obstacle to the realisation of participatory democracy is not the competence of citizens and the com-plexity of societal development, but is from the entrenched privileges of the political, technocratic, and economic elites.

Participatory democracy is a social process of optimising the breadth and depth of social participation in the realisation of shared ideas and ideals in everyday practices and strategies. Theories of participatory demo-cracy should attempt to inform this social process as a political process of deliberating the conditions for human freedom, as understood in the light of dialectical and hermeneutical reflections on human potential and its distortions (or suppressions). The task of theory is not to provide models for participation or a set of constitutional principles, but, rather, is to provide points of departure for critical reflections upon the conditions, possibilities, and limits of democratic participation, in order to recover the potential for critical examination of obstacles to and distortions of human freedom and rationality. Any theory of participatory democracy should be a political philosophy – aiming to inspire reflection and critical thought, even at the risk of being a work of fiction – quite opposed to the positivistic methods of political science. Rather than providing specific models for social organisation, theories of participatory democracy need to help us better understand how to satisfy the following five minimal conditions for the recovery of the practical value of democratic participation:

(1) *Effective participation*: Citizens need to have opportunities to develop the political skills needed to articulate concerns and preferences, to critically analyse their conceptual framework and participate in the setting of the public agenda, and learning how to better coordinate their efforts;

(2) *Enlightened understanding*: Citizens need to have opportunities to learn how to better critically understand their circumstances, gather information, identify their concerns and preferences, and for discovering common goods and a shared vision of the good life;

(3) *Political equality*: All concerns and preferences, as well as the proposed means to satisfy them, must be equally worthy of public deliberation and consideration – this does not mean that they are all equally acceptable or practical, but it does mean that they should be given a fair hearing, along with different perspectives on what qualifies as acceptable, practical, and a fair hearing;

(4) *Open and flexible agenda*: The process through which the agenda is set must be free from procedural and technocratic constraints – participants must be able to refine or re-evaluate the agenda at any stage;

(5) *Dynamic inclusiveness*: All concerned citizens must be included in deliberation and decision-making.

By satisfying these minimal conditions, participatory democracy promises to satisfy the following five important conditions for societal development:

(1) *Moral*: human beings are ends-in-themselves and have equal moral worth, the ideal society should be egalitarian;

(2) *Existential*: human dignity is bound up with authenticity and freedom, the ideal society should be libertarian;

(3) *Prudential*: human survival depends on reasoned deliberation and practical cooperation, the ideal society should be rational;

(4) *Ecological*: human survival depends on sustaining complex ecological conditions and relations, the ideal society should be sustainable;

(5) *Experimental*: all human beings are fallible and have limited knowledge and foresight; the ideal society should be open.

By satisfying these conditions, participatory democracy would maximise societal plurality and diversity, increasing the available societal stock of creativity, knowledge, experience, skills, and imagination required to compensate for human fallibility and uncertainty in an open-ended, changing, and complex world, by decentralising collective direct action and cooperation in accordance with principles of localisation and free-association, as dual aspects of maximal inclusion in the democratic construction of a rational society. However, if a democratic society is to avoid the potential pit falls of intellectual elitism and authoritarianism, the theoretical understanding of the possibilities and limits of democratic participation should be the outcome of a collective social effort to understand ourselves and the world within which we exist. As Dimitrios Roussopoulos and George Benello put it,

> It is appropriate that the first effort of conceptualizing the nature of participatory democracy should be a collective one. We need such intellectual task forces evaluating the work of contemporaries as well as the concrete experiences in the movement for radical social change... the debate goes beyond the concerns of liberalism, social democracy, and the authoritarian Old Left. The thrust... is to seek for structural changes in a system that is closing out true democracy and that is robbing people of the chance to have politically meaningful identities within their culture. All this is not simply a question of programs and ideologies, but of the prevailing institutions themselves.[26]

It is essential for the practical success of participatory democracy that it satisfies Galston's four conditions for liberal pluralism based on free-association:[27]

(1) *Knowledge*: an awareness that there exist alternative ways to live to the one that one is in fact living;
(2) *Capacity*: the ability to assess the alternatives if it becomes desirable to do so;
(3) *Psychological*: freedom from propaganda, brainwashing, and other kinds of psychological manipulation or coercion;
(4) *Fitness*: the ability of anyone to actually live an alternative way of life.

Hence, even though a participatory democracy would not be obliged to conform to Kropotkin's specific vision of an ideal community, it should share his general vision of the ideal society as being one that

> seeks the fullest development of free association in all its aspects, in all possible degrees, for all conceivable purposes: an ever changing

association bearing in itself the elements of its own duration, and taking on the forms which at any moment best correspond to the manifest endeavours of all... a society to which pre-established forms crystallised by law are repugnant, which looks for harmony in an ever-changing and fugitive equilibrium between a multitude of varied forces and influence of every kind, following their own course.[28]

Evolution or revolution?

One may readily agree with Lewis Mumford, that the ultimate test of the degree of any civilisation is not its scientific and technological capabilities, but is whether it uses these capabilities to enhance the potential for human flourishing and creativity.[29] It is possible that a small democratic society with a low level of industry can demonstrate a greater capacity for human flourishing and creativity, and, therefore, be more civilised, than a large empire with a massive productive base that enslaves the vast majority for the benefit of preserving great wealth and privilege for an economic elite. The ideal basis for the democratic development of a civilised society requires the conversion of increased efficiency and productivity of that society into increased time for leisure and creative activity for every member of that society. The modern representation of increased productivity and efficiency as a means to the further increase of productivity and efficiency demonstrates the extent that modern society has not even achieved the basic level of a comprehensive scheme of societal ends and a meaningful existence upon which a democratic civilisation would be founded. It is clear that the scientific and technological development of modern society would be more democratic if it were more accessible and developed by all citizens in accordance with their vision of how they should live. A genuine democratisation of technology would significantly empower women in all areas of economic, political, and social life. In a genuine democracy, it is incredibly important to allow children to participate in the structuration of their schools, play grounds, parks, and other public spaces, as part of their education is how to be a citizen. Of course, while technology clearly helps disabled people overcome physical difficulties, a complex technological infrastructure tends to discriminate against disabled people, especially when the work environment is increasingly becoming one within which all workers are being reduced to a set of simple physical operations and, thus, permits increased opportunities for the exploitation of disabled people, but, if disabled people are able to participate in the design of technological infrastructures and public spaces, then not only would this provide better access and improve the aesthetics of the lived-world for disabled people but, also, society would be enhanced by a significant increase in its available stock of knowledge, skills, ideas, creativity, and lateral thinking. Any society that marginalises significant proportions of its population reduces its potential for rational societal development. It is of great benefit for a society to allow all of its citizens to participate in its construction and development. By removing disadvantages and obstacles to universal participation in society, the society as a totality is empowered and enriched. If marginalised and frustrated citizens are involved in the process of deciding how best to remove those disadvantages and obstacles, participating in the process of securing their own liberation, then this will increase the pluralism of society and its capacity for diverse, creative, and democratic participation. Of course this is easier said than done. First it is necessary to identify and analyse the technological structures that disempower groups and resists change. In order to do that, it is essential to ask the members of the disempowered groups and help them experiment with their own ideas for change. Secondly, it is necessary to follow such experiments through and protect them from the structures that will resist and erode all efforts to change the *status quo*. Thirdly, it is necessary to integrate the experiments into society in general by developing and implementing protocols and channels of communication between democratic assemblies and other organisations, in order for local communities to request additional help or pass on information, but, in a democratic society this can readily be built into the system through the election of delegates and the democratic structuration of the organisation of communications, resources, expertise, information, and training. It also improves the capacity of local communities to develop, adapt to, and integrate technological innovations within the structure and organisation of that community, in accordance with the local democratic consensus regarding the needs and character of that community.

Local differences must be respected as sources of dissent, creativity, diversity, and pluralism, involving different moral and epistemological standards, in order to resist the tendency among social theorists to assert the universality of their own moral and epistemological standards as the basis for a universal theory of democracy, with the implication that those of us who think differently need to be corrected in order for our consciousness to be sufficiently prepared for proper political activity. It is for this reason that participatory democracy cannot become an ideology with a definitive doctrine, without undermining itself, but rather gives political form to decentralised grass-roots movements. New participatory institutions must be built by citizens for themselves, who will use them to empower themselves through decentralised alliances of political action groups and organisations, trade unions, NGOs, charities, churches, etc. The challenge facing such grass-roots movements is how to effectively organise collective action, without needing any centralised leadership, authority, or bureaucracy. This can be achieved through polyarchic alliances and the creation of majorities through the over-lapping concerns and interests of minorities. Understood in this way, participatory democracy involves, as Roussopoulos and Benello put it,

> ...a politics of creative disorder... at once orientated to unveiling the inequities of the present, and to building a counter-system that is participatory from the ground up. New participatory systems must be built in all social spheres and, as they develop, will claim legitimacy and recognition as being genuinely democratic and accountable to their constituencies.[30]

A participatory democracy would not be an *ochlocracy* (disorder of the rabble). It does not require a mass uprising to "smash the state" or "overthrow the government", given that these actions tend to result only in further violence and tyranny, but, rather, it is a gradual process wherein citizens gain the confidence and skills to better govern their own affairs and gain independence from centralised government. Once the state apparatus ceases to mediate social relations and practices, they become obsolete. A participatory democracy cannot occur while citizens are either dependent upon the State or refuse to take responsibility for societal development because the ideals of freedom and equality can only be realised through active participation in developing the economic and political conditions of human existence. These ideals cannot be achieved for the citizenry by a "revolutionary vanguard" or "founding fathers". Democratic emancipation requires the gradual dissolution of the state apparatus by a committed and active participation of the citizenry in their own emancipation through locally and directly resolving their practical affairs and concerns for themselves. This cannot be done for the citizenry because it requires the conscious formation of the *demos* as the sovereign power of free-association through democratic assembly, communication, coordination, and cooperation to discover and realise shared visions of the ideal society, the good life, and human well-being, as the result of maximal inclusion in localised experiments and traditions. The transition to participatory democracy requires the widespread development of citizens' commitment to democratic participation; public concern with the quality of communities and workplaces; civic virtue, social justice, equality, and personal responsibility; effective communications (wherein any citizen can talk to any other); specific places for citizens to assemble; practical processes of deliberation and decision-making; opportunities for public discussion on the nature of citizenship and the goals of democratic participation; cooperative activities within local communities and workplaces; free-associations and alliances between democratic assemblies; publicly owned and locally controlled media; publicly owned and locally controlled education; respect for human rights; democratisation of trade unions; alliances between political movements; international assemblies, conferences, and forums; and experiments in workplace and community democracies.

The *demos* could express its sovereign power in two aspects: the Great Refusal and the Turning Away. The Great Refusal involves the refusal on the part of the vast majority to participate in sustaining the structures of its own oppression. Through general strikes, civil disobedience, boycotts, and mass protests, non-violent resistance movements (such as those following the lead of Ghandi and Martin Luther King) can effectively achieve radical transformations in the popular awareness of the capacity for democratic participation and undermine the structures of oppression.[31] However, despite the inspirational and educational effect of popular awareness, this has only a negative – deconstructive – aspect of democratic participation. It creates genuine opportunities for the Turning Away of large numbers of people from the *status quo* and for the positive – constructive – aspect of democratic participation in building a new society. The Turning Away involves the development of positive community experiments in forming free-associations to administrate public affairs; forming cooperatives to coordinate means to achieve shared ends; working towards local independence and self-sufficiency; and, forming democratic community and workplace assemblies. The withering away of the State involves both the Great Refusal and the Turning Away, as a direct democratic and gradual social process of reducing dependency on the State and transferring the administrative functions of the State to civic society. It is essential, especially during the transition phase between liberal to participatory democracy, that polyarchic organisations confront the *status quo*, galvanising the call for democratic participation into a multi-faceted political movement to be taken very seriously indeed as a decentralised means of social organisation, eventually forming the civic structure for the administration and legislation of society. This involves directly challenging the legitimacy of the *status quo*, refusing to support it where and when it is unable to justify itself, and replacing it with alternative institutions and organisations. Given that a participatory democracy would threaten the privileges of technocrats and professional politicians, as well as their masters within the economic elite, the citizenry cannot expect to be "granted" the right to govern its own affairs, but must learn how to do so without waiting for permission. The process of transformation of society into a participatory democracy would be a gradual process, allowing people time to learn how to govern their own affairs, but this does not mean that the initiation of the transformation should be deferred until some future date when (supposedly) the conditions will be suitable (of their own accord). It requires a societal development of citizens' consciousness, ability, and confidence about the possibilities and potential of their participation in the public realm. The state apparatus needs to be absorbed by civic society and this can only occur in direct proportion to the willingness and ability of citizens to take responsibility for the governance of their own communities and workplaces. Furthermore, as history shows, radical and revolutionary changes are inherently unstable as a result of the historical discontinuities between the old and the new social order, and, the abruptness of such changes require a transition phase within which people can learn how to participate in the societal transformation through voluntary cooperation. Otherwise, any such revolution has a high risk of degenerating into barbarism or totalitarianism, followed by a return and reinforcement of the old social order, either because it is familiar or more tolerable than either barbarism or totalitarianism. Gradualism recognises that successful political revolutions are also social evolutions, and, therefore, only fully participatory and voluntary political revolutions can be successful. Historical political revolutions have required and recognised the need for volunteerism, but resorted to propaganda and coercion when it was not forthcoming. Such revolutions were premature and inevitably failed. A transition phase must reject any concept of "counter-revolution" and allow citizens time to come to understand themselves as the *demos*, allowing each and every citizen to incorporate social changes into their lives at their own pace and in their own way, allowing criticism of those revolutionary changes and also dissent from them if citizens are so inclined. The transition phase is a societal maturation and, therefore, all revolutionary efforts should be generated by citizens showing its local practical and emancipatory value to their neighbours and fellow workers. The initiation of participatory democracy occurs when decentralised groups of citizens (each trying to enrol their neighbours and fellow workers by showing the practical and emancipatory value of their ideas) freely associate with each other to form a loose confederation of small citizens' assemblies and direct action groups. The development of that confederation into an organised federation of local democratic assemblies and political movements would be a gradual process and constitute the political form of the transition phase – as citizens learn how to cooperate to organise production, distribution, communications, transportation, etc., involving agriculture, industry, services (including healthcare, childcare, education, care for the elderly and disabled), and media, as well as sanitation, etc. Given that all these activities are done by the citizens anyway, but instead of being motivated by wages, it is a matter of the citizenry learning to do these things for themselves, as cooperative endeavours, and to give them political form as instances of democratic participation in the administration of societal development. It has taken a very long time for the infrastructural networks of modern society to be developed to satisfy the needs of citizens. The transformation from hierarchical to democratic structures will not only be a matter of dissolving authority and removing privileges, but it will essentially be an education process wherein citizens learn political efficacy and social skills.

People need to recover their sovereign power by controlling local community development; rebuilding local food markets and trades; building and maintaining community infrastructure; developing cooperative enterprises; developing communications networks; forming neighbourhood and municipal assemblies; collectivisation of resources, skills, knowledge, and experience; achieving greater levels of representation in local government and more control over the local budget/taxes; and support efforts to localise economic development. Localisation has practical value because it reduces: fuel costs and dependency on oil; international conflicts and the need for maintaining a large professional army; international terrorism and the need for maintaining a police state; worker unrest and industrial action; trade deficits; alienation and extremes of poverty and wealth; and also reduces the local economic vulnerability to capital flight. Once citizens have organised all the practical aspects of their lives for themselves, having given them political form as examples of participatory democracy, citizens can organise policing, defence, and their legislation of their communities and workplaces. Once the administration of public affairs are organised through democratic assemblies, cooperatives, and other forms of free-association, people will be in a position to legislate their own affairs – directly transferring this function from the State too, by forming the Constitutional Assembly, via neighbourhood, municipal, and regional assemblies, wherein the citizenry can begin the process of deliberating and writing the Constitution, as an expression of how they understand democracy, rights, justice, duties, and citizenship, as a guide for learning what it is to be the *demos*. At which point, the state apparatus are obsolete and the public refusal to pay any taxes to the State or legitimate it via elections would deliver the *coup de grace*. The remnants of the State will crumble due to their impotence. It is only at this point should the citizenry take upon themselves the responsibility to be magistrates of the Constitution and directly enforce their own laws. By founding governance on interpersonal relations and direct action, as a thickening of thin democracy, social organisation can evolve into a participatory democracy based on decentralised, cooperative free-associations and alliances between them. A genuinely democratic revolution is not a process of gaining power over the State and using it to serve democratic decisions. This would increase the dependency upon the State, which would tend towards "democratic centralism", the subordination of the democratic process to technocratic constraints and demands, and would inevitably degenerate into an authoritarian technocracy. A genuinely democratic revolution requires independence from the State, achieved by a gradual and progressive social evolution. Hence, participatory democracy should emerge from the representative system, but it does so by making the representative system obsolete rather than abolishing or destroying it. In this respect, the democratic revolution would be synonymous with a social evolution. Self-governance involves people forgetting about the State by learning to live without relying on it, replacing all its functions by building polyarchic networks of citizens capable of sharing ideas, experiences, knowledge, and resources to build the infrastructure for their own communities based on mutual aid and over-lapping concerns and interests. Enrolment and persuasion, through free-associations and cooperative enterprises, rather than consensus and coercion, would constitute the democratic revolution as a decentralised mass movement of people taking responsibility for their own lives and helping each other to build a better society in accordance with their shared vision of what the ideal society should be. A genuinely democratic revolution (as opposed to a violent insurrection) presupposes the social evolution the shared vision of a better society among the vast majority and the practical capacity to realise it. In the absence of consensus regarding a shared vision, the critical and creative participation of all citizens in the deliberation and exploration of the conception of the rational society, in the context of the need for practical and cooperative action, would recover the social potential for the discovery of how to construct the rational society by recovering an awareness of how its contested nature creates the dynamic tension upon which its possibility as an open society depends. By opening rationality to question, in the absence of universally accepted epistemological and moral standards, the democratic process has the potential to maintain a sufficient degree of intellectual criticality and breadth (in equal proportion to the plurality and diversity of the *demos*) to hold genuinely alternative conceptions of rationality to be at stake, allowing all contenders to be equally worthy of consideration, without degenerating into relativism or nihilism. Hence, the outcome of the democratic process would not be a matter of imposing some consensus or collective will upon the world, given that the world does not conform to human intentionality. The democratic process would be a dialectical social ontology, which would be transformed as we respond to and accommodate the unforeseen consequences of our actions, as well as unexpected events and changes, while new possibilities arise, as our potentials are actualised and new goals and intentions (including new social needs) become available. Social being transforms human character and agency, given that we are in part defined by what we do and aspire towards, and it is in this sense that cooperative, democratic activities are transformative. Human history is the product of the dialectical movement between reasoned choices and their consequences, while the absence of absolute human control over this dialectical movement shows how human history and social evolution are brought together, as a combination of punctuated and gradual experimentalism, through criticality and creativity, in a complex, changing, and open-ended world that does not conform to human intentionality.

The democratisation of all the public institutions of society requires the direct participation of large numbers of people. As Roussopoulos and Benello noted,

> In order to realise such a vision we need a movement that lives it. The implications of building such a movement are revolutionary, for it confronts the basic institutional structures of the present society. Confrontation is not without implications of violence when demands for participation are met with repression; but it is also a non-violent revolution since its confrontation is on the basis of an alternative vision that seeks expression in counter-institutions that adequately

express the vision. Such counter-institutions can act, both as paradigms and experiments embodying a new vision, and as sustaining bases for continuing resistance and confrontation.[32]

Once we recognise that a genuinely democratic revolution presupposes a social evolution, emergent from successful grass-roots struggles and experiments in local community and workplace democracy, then we must also recognise that, without this social evolution, any political "revolution" will amount to little more than a *coup d'état* and, as can be seen from the examples of the French and Russian Revolutions, would likely degenerate rapidly into an intensification of hierarchical structures capable of extremes of terrorism and irrationality all in the name of preserving "the Revolution". Without a radical evolution in the consciousness and social organisation of the vast majority and the popular commitment to participatory democracy, any *coup d'état* performed in the name of "the People" by a "revolutionary vanguard" will necessarily be *for the people* rather than *by the people*. As a result, any revolutionary changes in social organisation, according to the theoretical vision of a minority, will necessarily be coercive and require a centralised authority, which, from the onset, suppresses any possibility of democratic participation on the part of the majority because it requires the construction of pervasive propaganda campaigns and police-state tactics to galvanise support for the ruling minority and suppress dissent among the majority. Perhaps as a matter of historical necessity, any "revolution" based on a *coup d'état* will result in the promotion of widespread irrationality and fear being the tools of maintaining centralised authority. The promotion of fear and irrationality undermines the societal capacity for the development of civic and intellectual virtues necessary for a genuinely democratic society. Democratic participation would involve a perpetual social evolution of the revolutionary struggle to liberate human potentiality, achieved through a dialectical recovery of its emancipatory ideal from the constraints of its actuality, as people learn their capabilities through experimentation, challenging their limitations, and exploring alternative possibilities for social organisation at a local level. Cooperative enterprises and free-associations would not only provide practical means to solve everyday problems, but they will also deepen the sense of respect and trust between citizens, help people learn how to coordinate their efforts to make better decisions in a complex, open-ended, and changing world, develop their communication skills, interpersonal intimacy, civic virtues, and the sense of fellowship required to transform a series of meetings into a dynamic, evolutionary process of the development and differentiation of the ontology of social being through voluntary participation in decentralised democratic experimentalism.

Ortega and Arendt's criticisms of "direct action" of the masses (for being acts of violence without standards of legitimacy based on reason or justice) do not apply to a participatory democracy. As a "thickening of thin democracy", it is also the case that the development of participatory democracy affirms Arendt's distinction between political action and social organisation. Reasoned deliberation is essential for the development and preservation of the conditions for freedom, without which, political action is reduced to coercive power or violence. The *modus operandi* of participatory democracy is to enhance the human capacity to exercise reason and justice, in the absence of universally agreed standards, as well as legitimate public participation in the equitable division of labour and distribution of resources, and, as such, it gives social organisation political form. Its appeal to egalitarianism and voluntarism are pragmatic response to epistemological and moral pluralism in the face of irresolvable ambiguity and uncertainty. In this respect, the principles of maximal inclusion, localisation, and free-association are pragmatic conditions for the possibility of recovering the practical value of reasoned deliberation regarding the best course of action. A high degree of societal tolerance to pluralism and diversity is a pragmatic condition for maximising the potential for the development of an open, rational, and sustainable society, in the absence of universally agreed epistemological and moral standards in a complex, open-ended, and changing world that does not always conform to human intentions and expectations. It becomes an ontological condition for the maximisation of the societal capacity for survival, creativity, experimentation, and freedom. This is a fundamentally important presumption for the reasonable expectation of the possibility of building local communities, as sites of social being, through cooperative endeavours based on reasoned deliberation, while maximising the practical value of criticism and dissent. Without this presumption, democracy could not amount to anything beyond the public expression of preferences and concerns; the assertion of absolute truths would degenerate into factionalism and the high likelihood of power struggles and even civil war. It is this presumption that provides a rational foundation to the intellectual openness to one's neighbour, as being someone equally worthy of consideration, meaningful, practical, and reasonable. As Ortega put it,

> Restrictions, standards, courtesy, indirect methods, justice, reason! Why were all these invented, why all these complications created?

They are summed up in the word civilisation, which, through the underlying notion of *civis*, the citizen, reveals its real origin. By means of all these, there is an attempt to make possible the city, the community, common life. Hence, if we look into all these constituents of civilisation just enumerated, we shall find the common basis. All, in fact, presuppose the radical progressive desire on the part of each individual to take others into consideration. Civilisation is before all, the will to live in common.[33]

For Ortega, the revolt of the masses – the construction of mass society – is an act of destructive violence against the conditions for civilisation in favour of a return to the jungle, to primitivism – barbarism. The revolt of the masses is a rebellion against the desire to seek some higher intellectual authority and submit to its discipline, be it reason, truth, justice, science, or art. However, once we recognise that the human condition involves diverse and pluralistic higher intellectual authorities, without any universal agreement on how to decide between them, then the liberal republicanism affirmed by Ortega is insufficient to adequately incorporate this level of pluralism and diversity, without distorting or suppressing it through the coercive power required to centralise and unify the State as the neutral site of the public realm. As a "thickening of thin democracy", a genuine democratic revolution is not a revolt at all, but a social evolution that takes communitarian republicanism, liberal democracy, and social libertarianism as the starting points in the development of its cultural, conceptual, and existential meaning, wherein each and every citizen is a nodal point of variation, selection, and experience in the ongoing differentiation and development of the particularities of community life and the generalities of society.

Theories of participatory democracy must appeal to reason and practicality, rather than revolutionary fervour and zeal. Hence, visions of the ideal society, if they are to avoid the charge of being utopian fictions must successfully appeal to reason and practicality to show that:

(1) the realisation of this vision is possible;
(2) the potential for this realisation can be recovered from the actuality of the *status quo*;
(3) this vision has a cultural connection with an historical tradition or movement;
(4) the realisation of this vision would improve the ability of human beings within particular communities and workplaces to satisfy or change the material conditions of their existence;

(5) practical experiments in economic and political relations can be derived or developed from this vision;

(6) there is a reasonable expectation that the realisation of this vision would resolve current struggles or crises;

(7) this vision does not presuppose its own universality, but, instead, recognises cultural and subjective differences will lead to dissent and criticism;

(8) the realisation of this vision does not overly rely on sustained good will, heroism, or sacrifice;

(9) there is a reasonable expectation that the realisation of this vision would be a societal good for the vast majority, if not everyone;

(10) the conditions for the realisation of this vision are themselves practical, reasonable, and desirable.

Historical examples show us that experiments in democratic participation fail due to the following reasons:

(1) External hostility and violence;

(2) Lack of privacy between members – leading to disputes and feuds;

(3) Lack of commitment to the success of the experiment;

(4) Insufficient knowledge, skills, and resources;

(5) Isolation from other communities – leading to a closed community;

(6) Lack of a clear vision;

(7) Factionalism and authoritarianism;

Taking these reasons into account, an adequate theory of participatory democracy must include:

(1) Strategies and tactics for collective self-defence;

(2) An account of how the boundary between the public and private should be negotiated and respected;

(3) A recognition that a shared commitment between members to the success of democracy is a fundamental prerequisite – every democratic process will fail without it;

(4) An account of how knowledge, skills, and resources should be acquired and distributed throughout the community or workplace;

(5) An account of how communication and negotiation between communities and workplaces should be conducted;

(6) An account of how a shared vision can be discovered and realised through the democratic process;

(7) An account of how pluralism, diversity, and dissent should be incorporated into the democratic process.

As the examples of Soviet repression of dissent and democratic movements in Hungary (1956), Czechoslovakia (1968), and Poland (1981) show, such violent and oppressive acts were only effective in propping up illegitimate regimes in the short-term by suppressing and delaying the emergence of participatory democratic movements. The events in Eastern Europe between 1988 and 1990 showed that even very oppressive regimes were unable to indefinitely contain democratic resistance which grew to such proportions that it ultimately undermined the sustainability of the Soviet Union. The subsequent superposition of "the victory of the West" over the democratic *glasnost* within Russia and Eastern Europe is little more than ideological propaganda that fails to acknowledge how participatory movements transcended the limitations of Gorbachev's *perestroika*. The subsequent imposition of neoliberal market reforms and the liberal model of representative democracy over Eastern Europe and Russia was the result of the repression of those calls for democratic participation – freedom of expression and association – in order to impose the conditions for the economic exploitation of these countries. The emergence of new social possibilities were shut down in order to transform Soviet state-capitalism into markets available to multinational corporations, while the remnants of the old political elite positioned themselves into the positions of power within the new "democracies" in order to reap the benefits of the transition and gain private ownership of former national industries and land.[34] This was not an ideological victory of Western liberal democracy, as Francis Fukuyama and others have asserted.[35] Rather than resulting in the culmination of humanity's "ideological evolution", this simply imposed neoliberal economic policies and the competitive elitism model as the form of government. Rather than liberate the citizenry, this has resulted in the domination of their markets by multinational corporations, an increasing gap between economic classes, leading to intensified control over the natural resources, massive unemployment, the further erosion of the middle-class, an intensification of corruption and economic slavery, increases in crime, emigration of skilled and educated citizens, the erosion of public healthcare and education, and the further erosion of governmental legitimacy. Far from being "the end of history", the so-called "victory of the West" has resulted in an intensified class-struggle, wherein resistance and opposition to neoliberal exploitation and imperialism has become a matter of survival for the majority of human beings *qua* human beings. This struggle is not only the result of the inability of "the-have-nots" to accept and accommodate themselves to the new world order, but is an ideological struggle for democracy and social justice in a world wherein the vast majority are increasingly oppressed by the "minority rule" of the economic elite that are seemingly constructing a global neo-feudal system. While Fukuyama's point about the exhaustion of "grand narratives" is perhaps poignant, in a postmodern sense, it not only presupposes that consumerism has trumped idealism in the West, but also has neglected the significance of rising nationalism and religious fundamentalism in opposition to the new world order. It seems that his celebration of "the victory of the West" was premature and we are still very much the subjects and objects of history.

Notes and References

Chapter 1   The Call for Democratic Participation

1. Andrews, G.R. and Chapman, H. (eds), *The Social Construction of Democracy 1870–1990* (Basingstoke: Palgrave Macmillan, 1995).
2. Feenberg, A., *Critical Theory of Technology* (New York: Oxford University Press, 1991).
3. Winner, L., *Autonomous Technology: Technics-Out-of-Control as a Theme in Political Thought* (Cambridge, Mass: MIT Press, 1977).
4. Winner, L., *The Whale and the Reactor: A Search for Limits in an Age of High Technology* (University of Chicago Press, 1986).
5. For example see Winner's introduction in Winner, L. (ed.), *Democracy in a Technological Society* (Dordrecht: Kluwer, 1992).
6. For discussions of implicit gender prejudices in scientific methods and theories see Haraway, D.J., *Primate Visions: Gender, Race, and Nature in the World of Modern Science* (New York: Routledge, 1989). See also Gattiker, U.E. (ed.), *Women and Technology* (New York: de Gruyter, 1994); Harding, J. (ed.), *Perspectives on Gender and Science* (London: Farmer Press, 1986); Easlea, B., *Fathering the Unthinkable: Masculinity, Scientists, and the Nuclear Arms Race* (London: Pluto Press, 1983); Rothschild, J. (ed.), *Machina Ex Dea: Feminist Perspectives on Technology* (New York: Pergamon Press, 1983); Schwartz Cohen, R., *More Work For Mother* (New York: Basic Books, 1983); Merchant, C., *The Death of Nature: Women, Ecology and the Scientific Revolution* (New York: Harper & Row, 1980).
7. See Daniel Kleinman's introduction in Kleinman, D.L. (ed.), *Science, Technology & Democracy* (New York: State University of New York Press, 2000).
8. Rogers, K., *Modern Science and the Capriciousness of Nature* (Basingstoke: Palgrave Macmillan, 2006), Chapters 5 and 6.
9. Roszak, T., *Where the Wasteland Ends* (Garden City, NY: Doubleday, 1972), p. 400.
10. Rogers, K., *On the Metaphysics of Experimental Physics* (Basingstoke: Palgrave Macmillan, 2005).
11. Pickering, A., *The Mangle of Practice* (University of Chicago Press, 1995). See *On the Metaphysics of Experimental Physics*, Chapter 6, for a critical discussion of the applicability of Pickering's thesis to experimental physics.
12. See *On the Metaphysics of Experimental Physics* for further discussion.
13. Held, D., *Models of Democracy* (Stanford University Press, 1987).
14. Lippman, W., *The Phantom Public* (New York: Harcourt Brace, 1925).
15. Weber, M., *The Protestant Ethic and the Spirit of Capitalism* (Parsons, trans., New York: Scribners, 1958) and *Economy and Society* (2 vols., Berkeley, CA: University of California Press, 1978).
16. Weber, M., "Politics as a Vocation" (in Gerth and Mills (ed.), *From Max Weber*, New York: Oxford University Press, 1972).

220

17. Schumpeter, J., *Capitalism, Socialism and Democracy* (New York: Harper, 1950), first published in 1943.
18. *ibid*, p. 272.
19. *ibid*, pp. 284–5.
20. See *ibid*, p. 251 and pp. 252ff.
21. *ibid*, pp. 253–68.
22. Pateman, C., *The Problem of Political Obligation: A Critique of Liberal Theory* (Cambridge: Polity Press, 1985).
23. This has been widely known by political theorists since Edward Bernays published *Propaganda* (New York: Liveright, 1928), explaining how commercial advertising and mass media techniques could be used for political purposes to gain public support for policy and candidates. For interesting and detailed contemporary discussions see Ellul, J., *Propaganda: The Formation of Men's Attitudes* (New York: Alfred Knopf, 1965); Herman, E.S. and Chomsky, N., *Manufacturing Consent: The Political Economy of Mass Media* (New York: Pantheon Books, 1988); Chomsky, N., *Media Control: The Spectacular Achievements of Propaganda* (New York: Seven Stories Press, 1997); Sproule, M., *Propaganda and Democracy: The American Experience of Media and Mass Persuasion* (Cambridge University Press, 1997). Barsamian, D. and Chomsky, N., *Propaganda and the Public Mind* (Cambridge, Mass: Southend Press, 2002); Hightower, J., *Cable News Confidential* (Sausalito, CA: Polipoint Press, 2006).
24. See *CSD*, pp. 256–64 for examples of Schumpeter's low opinion of the intellectual and moral capacity of the majority of people. For an interesting critique of how Schumpeter's prejudices underwrote his objections to direct democracy, see Parry, G., *Political Elites* (London: Allen and Unwin, 1969).
25. *ibid*, pp. 295–6.
26. Berlin, I., "Two Concepts of Liberty" in *Liberty* (New York: Oxford, 1958).
27. Galston, W., *Liberal Pluralism: The Implications of Value Pluralism for Political Theory and Practice* (Cambridge University Press, 2002). See also Kymlicka, W., *Liberalism, Community, and Culture* (Oxford University Press, 1989).
28. *MSCN*, pp. 178–88.
29. For detailed discussions see Landes, D.S., *The Unbound Prometheus: Technological Change and Industrial Development in Western Europe from 1750 to the Present* (Cambridge University Press, 1969); Goldthorpe, J.H. (ed.), *Order and Conflict in Contemporary Capitalism* (Oxford University Press, 1884); Sklar, M.J., *The Corporate Reconstruction of American Capitalism, 1890–1916: The Market, Law, and Politics* (Cambridge University Press, 1988); Chandler, A.D., *Scale and Scope: The Dynamics of Industrial Capitalism* (Cambridge, Mass: Harvard University Press, 1990).
30. Prezeworski, A., *et al.*, *Democracy and Development: Political Institutions and Well Being in the World* 1950–1990 (Cambridge University Press, 2000).
31. Denis Mueller, and Thomas Stratmann, in "The Economic Effects of Democratic Participation" (in *Journal of Public Economics*, 87, 2003) studied thirty-eight "representative democracies" from around the world and concluded that there was a tendency for high levels of political participation among the poorest citizens during elections to result in progressive economic policies and greater economic equality (in terms of a more equitable distribution of economic benefits). Hence, Mueller and Stratmann argued that this high level of correlation between participation and the more equitable

redistribution of wealth resulting from economic growth was a major factor in the motivation for the economic elite to use its political influence to disenfranchise poor citizens from the electoral process and public participation in general by means of media propaganda, voter suppression and intimidation tactics, and electoral fraud.

32. Lowi, T., *The End of Liberalism* (New York: Norton, 1969), and Habermas, J., *The Structural Transformation of the Public Sphere: An Inquiry into a Category of Bourgeois Society* (Burger and Lawrence, trans., Cambridge: MIT Press, 1991, first published in 1969). See also Miliband, R., *The State in Capitalist Society* (London: Weidenfeld and Nicolson, 1969); Keohane, R.O. and Nye, J.S. (eds), *Transnational Relations and World Politics* (Cambridge, MA: Harvard University Press, 1972); Held, D. *et al.* (eds), *The State in Capitalist Europe* (London: Allen and Unwin, 1984); Jessop, B., *State Theory* (Cambridge: Polity Press, 1990); Korten, D.C., *When Corporations Rule the World* (San Francisco: Berrett-Koehler, 1995); Crenson, M.A. and Ginsberg, B., *Downsizing Democracy: How America Sidelined its Citizens and Privatized its Public* (Baltimore: John Hopkins University Press, 2002); Drutman, L. and Cray, C., *The People's Business: Controlling Corporations and Restoring Democracy* (San Francisco: Berrett-Koehler, 2004); Smith, G.W., *The Politics of Deceit* (Hoboken, NJ: Wiley & Sons, 2004).

33. Also see *MEP* for a detailed discussion of this point.

34. In Thucydides' *The History of the Peloponnesian War* (Hobbes, trans., Chicago University Press, 1989), *techne* (artifice) is represented as the means by which human beings can overcome natural catastrophes and struggle against the harshness of Nature. See *MEP*, Chapter 3, and *MSCN*, Chapters 2 and 4, for discussions of Baconian science.

35. In 1961, in his farewell presidential address, Dwight Eisenhower warned that the state supported industrial-military complex could have damaging consequences for the US economy and democracy. For detailed discussions of these damaging consequences see Mumford, L., *Pentagon of Power: The Myth of the Machine* (New York: Harcourt Brace Jovanovich, 1970); Melman, S., *Pentagon Capitalism: The Political Economy of War* (San Francisco: McGraw-Hill, 1970) and *The Permanent War Economy: American Capitalism in Decline* (New York: Simon and Schuster, 1985); Chomsky, N., *Failed States: The Abuse of Power and the Assault on Democracy* (New York: Metropolitan Books, 2006).

36. For wide-ranging and detailed discussions see Callon, M., Law, J. and Rip, A. (eds), *Mapping the Dynamics of Science and Technology: Sociology of Science in the Real World* (London: Macmillan, 1986); Marcus, G.E. (ed.), *Techno-scientific Imaginaries* (University of Chicago Press, 1986); Aronowitz, S., *et al.* (eds), *Technoscience and Cyberculture* (London: Routledge, 1996); Downey, G.L. and Dumit, J. (eds), *Cyborgs & Citadels: Anthropological Interventions in Emerging Sciences and Technologies* (Santa Fe, NM: School of American Research Press, 1998); Gordo-Lopez, A.G. and Parker, I. (eds), *Cyberpsychology* (London: Macmillan Press, 1999)

37. See *MSCN*, Chapter 4.

38. Bimber, B., "Three Faces of Technological Determinism", Smith, M.R., and Marx, L. (eds), *Does Technology Drive History? The Dilemma of Technological Determinism* (Cambridge, Mass: MIT Press, 1994), pp. 80–100.

39. Mitcham, C., *Thinking Through Technology: The Path Between Engineering and Philosophy* (University of Chicago Press, 1994).
40. Dewey, J., *The Quest for Certainty* (New York: Capricorn, 1960), pp. 6–7.
41. Strauss, L., *The City and Man* (University of Chicago Press, 1964) and *Natural Right and History* (University of Chicago Press, 1953).
42. Gadamer, H-G., *Truth and Method* (London: Sheed and Ward, 1975).
43. Popper, K., *The Poverty of Historicism* (London: Routledge and Kegan Paul, 1957), Chapter 3.
44. For examples see Wright, S., *Molecular Politics: Developing American and British Regulatory Policy for Genetic Engineering, 1972–1982* (University of Chicago Press, 1994); Goggin, M. (ed.), *Governing Science and Technology in a Democracy* (Knoxville, TN.: University of Tennessee Press, 1986); Petersen, J. (ed.), *Citizen Participation in Science Policy* (Amherst, Mass: University of Massachusetts Press, 1984); Dickson, D., *The New Politics of Science* (University of Chicago Press, 1988).

Chapter 2   Substantive Theories of Technology

1. In *Ode To Man* in *Antigone* (New York: Dover Publications, 1993).
2. Mitcham, C., *Thinking Through Technology: The Path Between Engineering and Philosophy* (University of Chicago Press, 1994).
3. Weber, M., *The Protestant Ethic and the Spirit of Capitalism* (Parsons, trans., New York: Scribners, 1958) and *Economy and Society* (2 vols., Berkeley, CA: University of California Press, 1978).
4. Weber, M., "Science as a Vocation" (in Gerth and Mills (ed.), *From Max Weber*, New York: Oxford University Press, 1972).
5. Ellul, J., *The Political Illusions* (New York: Knopf, 1968).
6. See *MEP* and *MSCN* for detailed historical and philosophical discussions and references.
7. Winner, *Autonomous Technology*, p. 229.
8. Manheim, K., *Man and Society in an Age of Reconstruction* (New York: Harcourt, Brace & World, 1951), p. 55.
9. In the writings of Proudhon and Bakunin a libertarian, egalitarian, and rational society was inherently a post-scarcity society. Scarcity generates the fundamental struggle between human beings and the natural world and, which had been distorted into a struggle between human beings, therefore, science and technology were the means for human liberation. Technological determinism underscored the conditions for the transition to a communist society in the writings of Marx and Engels, wherein it is necessary that human beings are liberated from toil by science and technology, while being able to satisfy all the material conditions for human existence. Contemporary developments in anarchism and Marxism continue this traditional assumption of the emancipatory potential of science and technology. This position is a dominant theme in the social libertarian writings of Murray Bookchin. For example, see Bookchin, M., *Post-scarcity Anarchism* (New York: Black Rose, 1971); *The Philosophy of Social Ecology: Essays on Dialectical Naturalism* (New York: Black Rose, 1996), and *The Ecology of Freedom* (Oakland, CA: AK Press, 2005). Also see Biehl, J., *The Politics of Social Ecology:*

Libertarian Municipalism (New York: Black Rose, 1998). It is also dominant in the school of critical realism, as developed by Roy Bhaskar in *Science and Human Emancipation* (London: Verso, 1986), *The Possibility of Naturalism* (London: Verso, 1989), and *Dialectic: The Pulse of Freedom* (London: Verso, 1993).

10. *Capital* (Vol. 1, p. 352, note 2).
11. For Marx, there was an "absolute contradiction" between the technical demands of industrialisation and industrial capitalism (see *Capital*, I., pp. 533–4, for example). See also *MSCN*, Chapter 3, for a discussion of Marx's interpretation of the contradiction between the capitalist economic imperative and the technological imperative, wherein capitalism distorts or perverts the emancipatory essence of technology, which leads to structural contradictions between "the forces of production" and "the relations of production".
12. 1859 Preface to *Critique of Political Economy*.
13. See Blackledge, P. and Kirkpatrick, G., *Historical Materialism and Social Evolution* (Basingstoke: Palgrave Macmillan, 2002) for a detailed discussion of this point.
14. Arguably, Marx used the term "dictatorship" in the Roman sense of the term, as a temporary dictator who could save the Republic in terms of crisis, who had strictly limited constitutional powers, rather than the modern sense of the term. If this is the case, "the dictatorship of the proletariat" referred to a temporary period wherein productive workers would decide societal development for themselves and on behalf of the unproductive members of society.
15. Arendt, H., *The Origins of Totalitarianism* (New York: Meridian Books, 1958). See also Alexander Solzhenitsyn's *The Gulag Archipelago: An Experiment in Literary Investigation* (Whitney, trans., London: Harper & Row, 1973, published in the Soviet Union in 1989).
16. For an interesting distinction between the "critical" and "scientific" aspects of Marx's theory see Gouldner, A.W., *The Two Marxisms* (New York: Seabury, 1982). My criticisms of the inherent technological determinism within Marxism are directed towards the "scientific" aspects of Marx's theory, which have been dominant in the development of Marxism in theory and practice. If we take the "critical" aspects into account, as Andrew Feenberg has done, we move towards a substantive and critical theory of technology. I shall discuss this in the next chapter.
17. For further discussion of this point see *MSCN*, pp. 80–93.
18. Marx's conception of "the dictatorship of the proletariat" in a post-capitalist society was not that of a coercive state apparatus separate and above the proletariat. It was the democratic expression of the consensus of the producers of societal wealth, as a free and equal association of disciplined industrial workers who would collectively determine the division of labour, the forces and relations of production, and how to satisfy social needs, on the basis of technically rational decisions. See *Critique of the Gotha Program*; *Poverty of Philosophy*; and *Capital I*, *chap. 1*, *sec 4*, for example. According to Marx's theory, Russia was too industrially backward for a socialist revolution and the construction of the post-capitalist state. The Russian situation actually led Marx to reconsider his theory of the conditions for a socialist

revolution and the construction of a post-capitalist state to also include predominantly agricultural economies. Inspired by Georgy Plekhanov, Marx considered the Russian *mir* (traditional village based commune) to be the basic social unit for communal living, collective agriculture, and the fulcrum for social regeneration in Russia. With the aid of modern technology, such as agricultural machinery and chemical fertilisers, the *mir* could provide the foundational social unit for a socialist revolution in Russia, without needing to develop an advanced and disciplined industrial proletariat through capitalism first. See Shanin, T., *Late Marx and the Russian Road: Marx and the 'Peripheries of Capitalism'* (London, 1983). A similar theory was proposed by the nineteenth century social libertarian Alexander Herzen in *From the Other Shore and The Russian People and Socialism* (Budberg, trans., London, 1956). In some respects, this path was taken when Lenin and the Bolsheviks adopted the New Economic Policy in 1921 and permitted the development of an agricultural cooperative economy based around the *mir* and a mixed urban economy until Lenin's death in 1924. After its architects, such as Bukharin, had been undermined by Trotsky and the Left Opposition and the poor harvest in 1929, the conditions were right for Stalin to impose centralised collectivisation.

19. For descriptions of the distortion and suppression of science and technology in the Soviet Union see Zaleski, E., *et al.*, *Science Policy in the USSR* (Pairs: OECD Publications, 1969); Joravsky, D., *The Lysenko Affair* (University of Chicago Press, 1970); Bailes, K., *Technology and Society Under Lenin and Stalin* (Princeton University Press, 1978); Graham, L.R., *What Have We Learned About Science and Technology from the Russian Experience?* (Stanford University Press, 1988); Gorokhov, V., "Politics, Progress, and Engineering: Technical Professionals in Russia" in Winner, L. (ed.), *Democracy in a Technological Society* (Dordrecht: Kluwer Academic Publishers, 1992), pp. 175–86; and, Roll-Hansen, N., *The Lysenko Effect: The Politics of Science* (New York: Humanity Books, 2005).

20. As pointed out at the time by Rosa Luxemburg in *The Russian Revolution* (London: Paul Levi, 1922, written 1918), Rhys Williams, A., *Through the Russian Revolution* (New York: Bons and Liveright, 1921), and Emma Goldman in *My Disillusionment in Russia* (New York: Doubleday, Page & Company, 1923). Even John Reed who at the time of writing *Ten Days that Shook the World* (Penguin Books, 1966, first published in 1919) was both sympathetic and optimistic about the Bolsheviks as a democratic movement for genuine social revolution and justice, observed that the Bolshevik seizure of power was widely resisted and opposed among workers and peasants. Louise Bryant observed, in *Six Red Months in Russia* (London: Heinemann, 1919), few supporters for the Bolsheviks among the peasantry (which comprised about 80% of the population), who were largely disappointed by the Kerensky Provisional Government's failure to implement the promised land-reforms, but the peasants predominantly supported the Left Socialist Revolutionary Party, rather than the Bolsheviks. David Mandel, in *The Petrograd Workers and the Soviet Seizure of Power* (Basingstoke: Palgrave Macmillan, 1984), has criticised the claims made by many Western historians that the Bolshevik seizure of power was an example of an anarchic, opportunistic democratic workers' uprising. The Petrograd factory workers were a minority, even in Petrograd, and many of them disagreed with the Bolsheviks, many of whom

also disagreed with Lenin's call for a seizure of power. Mandel argued that the workers' soviets only sought workers' control over the factories, but did not seek control over the State. The Petrograd soviets only supported the Bolsheviks' seizure of power after it was a *fait accompli*. The Bolshevik seizure of power was not the result of a workers' uprising, but the workers' support for it was based on the need for a stable and practical government to deal with the political and economic circumstances created by the Kerensky Provisional Government's failures (including its violent suppression of the Bolsheviks, radical socialists, and the workers' soviets since April); its refusal to fulfil its promises to create a democratic Constitutional Assembly wherein the workers' soviets would share power with other democratic groups, such as political parties, municipal councils, trade unions, and the delegates from peasants' cooperatives; and, the deprivations caused by Russia's involvement in the First World War.

21. See Carr, E.H., *The Bolshevik Revolution 1917–1923, Volume 1* (Baltimore: Penguin Books, 1966); Althusser, L., *Lenin and Philosophy* (New York: Monthly Review Press, 1971); Lewin, M., *Lenin's Last Struggle* (London: Pluto Press, 1975); Liebman, M., *Leninism under Lenin* (London: Jonathon Cape, 1975); Bellis, P., *Marxism and the USSR: The Theory of Proletarian Dictatorship and the Marxist Analysis of Soviet Society* (Atlantic Highlands, NJ: Humanities Press, 1979); Fitzpatrick, S., *The Russian Revolution* (Oxford University Press, 1994); Gooding, J., *Socialism in Russia: Lenin and his Legacy, 1890–1991* (Basingstoke: Palgrave Macmillan, 2002) for detailed discussions of this point. See also Marcuse, H., *Soviet Marxism: A Critical Analysis* (New York: Columbia University Press, 1958). Marcuse's critical analysis of the contradictory tendencies in the Soviet Union analysed the fundamental ambivalence that was inherent to the attempt to realise Marxism, as an emancipatory and progressive aspiration for human freedom, happiness, and social justice, through suppressive and oppressive means that failed to realise any of the Marxist ideals in practice. Marcuse saw the structures and operations involved in the construction of the Soviet state apparatus as being essentially a series of responses to the West, he identified "objective tendencies" in these structures and operations that potentially could lead to liberating reforms as the Soviet Union developed into a technological society. He claimed that these "objective tendencies" were apparent in Khrushchev's reforms, but they were suppressed by the continued Stalinist objective of strengthening the State in response to the aggressive and interventionist policies of the Kennedy Administration. For a critical discussion of Marcuse's efforts to distance Marx's theory from Soviet Marxism, see Kellner, D., *Herbert Marcuse and the Crisis of Marxism* (London: Macmillan, 1984).

22. For an interesting and detailed discussion of how Stalin's brutal agricultural collectivisation policy was driven by the needs of rapid industrialisation, see Davis, R.W. *The Socialist Offensive: The Collectivisation of Soviet Agriculture 1929–1930* (Basingstoke: Palgrave Macmillan, 1980). See also Lewin, M., *Political Undercurrents in Soviet Economic Debates: From Bukharin to the Modern Reformers* (London: Free Press, 1975) and Sutela, P., *Economic Thought and Economic Reform in the Soviet Union* (Cambridge University Press, 1991).

23. Heilbroner, R.L., "Do Machines Make History?" first published in *Technology and Culture*, July 1967, pp. 335–45; reproduced in Smith, M.R., and Marx, L.

(eds), *Does Technology Drive History? The Dilemma of Technological Determinism* (Cambridge, Mass: MIT Press, 1994), pp. 53–65.

24. Bimber, B., "Three Faces of Technological Determinism", Smith and Marx (eds), *Does Technology Drive History?* pp. 80–100.

25. Reprinted in Rousseau, J-J., *The Basic Political Writings* (Indianapolis: Hackett, 1987), pp. 1–24.

26. Husserl, E., *The Crisis of European Sciences* (Carr, trans., Northwestern University Press, 1970), first published in 1910.

27. Heidegger, M., *Being and Time* (Macquarie and Robinson, trans., New York: Harper & Row, 1962).

28. For detailed discussion of Heidegger's criticisms of Galilean physics see *MEP*, Chapters 2 and 3.

29. Heidegger, M., "The Question Concerning Technology" (first published in 1955) in *The Question Concerning Technology and Other Essays* (Lovitt, trans., Harper Torchbooks, 1977) and *Basic Writings* (Farrell Krell, ed., London: Routledge, 1978, pp. 311–41).

30. "Letter On Humanism" (first published 1946) in *Basic Writings* (pp. 217–65), p. 255.

31. *Techne* (plural *technai*) had the loose meaning of art, craft, or science in ordinary Ancient Greek. It was used to convey craftiness, artfulness, or cunning in a device or ploy. It was only in the philosophies of Plato and Aristotle that *techne* was formally elevated to the status of the knowledge of the reasons or causal principles involved in productive or practical activity. See *MEP* for further discussion.

32. *QCT*, p. 15.

33. *ibid*, p. 25.

34. *ibid*, p. 32.

35. *ibid*, pp. 14 and 21–3.

36. For further discussion of this point see *MEP*, Chapters 3 and 4.

37. "Letter on Humanism", p. 259, footnote; see *MEP* for further discussion.

38. Marcuse, H., *Reason and Revolution* (Boston: Beacon Press, 1960), first published in 1941; *Eros and Civilization* (Boston: Beacon Press, 1966), first published in 1955.

39. Marcuse, H., *One-Dimensional Man* (Boston: Beacon Press, 1991), first published in 1964.

40. Marcuse was equally damming of both Western capitalist and "Soviet Marxist" state-controlled industrial societies. He was highly critical of the Soviet Union, see *Soviet Marxism* (New York: Columbia University, 1958), but refused to uncritically accept the capitalist propaganda regarding its moral superiority over communism.

41. See *MSCN*, pp. 93–104 for critical discussion.

42. See Marcuse, H., *Counter-Revolution and Revolt* (Boston: Beacon Press, 1972), and *The Aesthetic Dimension* (Boston: Beacon Press, 1978).

43. Ellul, J., *The Technological Society* (Wilkinson, trans., New York: Knopf, 1964).

44. For further discussion see *MSCN*, Chapter 3.

45. *Technological Society*, pp. 38–45. For critical discussion see *MEP*, Chapters 1 and 6.

46. For arguments in support of this claim see *MEP*, Chapter 4.

47. *TS*, p. 86.

48. *ibid*, p. xxxiii.
49. Arguably, the Christian spiritualism advocated by Ellul shows that he was something of an optimist. For details see his *The Politics of God and the Politics of Man* (Grand Rapids, Mich.: Eerdmans, 1972) and *The Ethics of Freedom* (Grand Rapids, Mich.: Eerdmans, 1976). See also Hanks, J.M., *Jacques Ellul: A Comprehensive Bibliography* (Greenwich, Con: JAI Press, 1984), and Lovekin, D., *Technique, Discourse, and Consciousness: An Introduction to the Philosophy of Jacque Ellul* (Bethlehem, Pa: Lehigh University Press, 1991).
50. *TS*, p. 184.
51. See Ellul, J., *The Technological Bluff* (Bomiley, trans., Grand Rapids, Mich: Eerdmans, 1990) for further discussion; see also *MSCN*, pp. 74–93.
52. *The Revolt of the Masses* (New York: Norton, 1993), first published in 1930.
53. Borgmann, A., *Technology and the Character of Contemporary Life: A Philosophical Inquiry* (University of Chicago Press, 1984) For critical discussions of Borgmann's philosophy of technology see Higgs, E., Light, A., and Strong, D. (eds), *Technology and the Good Life?* (University of Chicago Press, 2000), including Borgmann's "Reply To My Critics", pp. 341–70.
54. *Technology and the Character of Contemporary Life*, p. 3.
55. For a discussion of Borgmann's Aristotelian conception of character and virtue see Mitcham, C., "On Character and Technology" in Higgs, *et al. (ed.)*, *Technology and the Good Life?* pp. 126–48.
56. *Op cit*, p. 197.
57. *ibid*, p. 219.
58. See also Borgmann, A., "Communities of Celebration: Technology and Public Life" (in Ferré, ed., *Research in Philosophy and Technology*, vol. 10, Greenwich: JAI Press, 1990, pp. 315–45).
59. *Op cit*, p. 208.
60. *ibid*, pp. 223–31.
61. *ibid*, pp. 9–11.
62. For a detailed discussion of this point see Strong, D. and Higgs, E., "Borgmann's Philosophy of Technology" in Higgs, *TGL?*, pp. 19–37.
63. *Op cit*, pp. 35–48.
64. *ibid*, p. 53.
65. *ibid*, p. 105.
66. *ibid*, p. 198.
67. *MSCN*, pp. 154–6.

Chapter 3   Democratising the Technological Society

1. For example, see "Technological Determinism Revisited" in Smith and Marx, eds, pp. 67–78.
2. Feenberg, A., *Critical Theory of Technology* (Oxford University Press, 1992); *Alternative Modernity: The Technical Turn in Philosophy and Social Theory* (Berkeley: University of California Press, 1995); *Questioning Technology* (New York: Routledge, 1999); and *Transforming Technology: A Critical Theory Revisited* (Oxford University Press, 2002).
3. Feenberg drew heavily from the studies and analysis provided by Bijker. W., Hughes, T. and Pinch, T. (eds), *The Social Construction of Technological Systems:*

New Directions in the Sociology and History of Science and Technology (Cambridge, MA.: MIT Press, 1989) and Bijker, W. and Law, J. (eds), *Shaping Technology, Building Society: Studies in Sociotechnical Change* (Cambridge, Mass: MIT Press, 1992). See also Marx, L., *The Machine in the Garden* (Oxford University Press, 1964); Hughes, T.P., *Networks of Power: Electrification in Western Society, 1880–1930* (Baltimore: John Hopkins University Press, 1983); Callon, M., Law, J. and Rip, A. (eds), *Mapping the Dynamics of Science and Technology: Sociology of Science in the Real World* (London: Macmillan, 1986); Law, J. (ed.), *A Sociology of Monsters: Essays on Power, Technology and Domination* (London: Routledge, 1991); Smith, M.R. and Marx, L. (eds), *Does Technology Drive History? The Dilemma of Technological Determinism* (Cambridge, Mass: MIT Press, 1994); and Hughes, T.P., *Human Built World: How to Think about Technology and Culture* (University of Chicago, 2004). For an interesting discussion and development of Feenberg's sociological critique of technological determinism see Jacobsen, J.K., *Technical Fouls: Democratic Dilemmas and Technological Change* (Oxford: Westview Press, 2000).
4. Karin Knorr-Cetina argued that understanding the transcientific criteria imposed on scientific research is essential for understanding scientific judgements. See *The Manufacture of Knowledge* (Pergamon, 1981).
5. See also Ihde, D., *Technology and the Lifeworld* (Bloomington and Indianapolis: Indiana University Press, 1990), pp. 128–44.
6. *Questioning Technology*, pp. 207–10; see also Hans Jonas' Philosophical *Essays: From Ancient Creed to Technological Man* (Englewood Cliffs, N.J.: Prentice-Hall, 1974) and *The Imperative of Responsibility: In Search of an Ethics for the Technological Age* (Chicago University Press, 1984).
7. Parallels with Michel Foucault's analysis in *Knowledge/Power: Selected Interviews and Other Writings, 1972–1977* (Gordon, trans., ed., New York: Pantheon Books, 1980). As sociologists have argued for many decades, scientists do not need to be involved in any conspiracy to serve the interests of social power, but simply do so by neglecting to question matters outside of their areas of specialisation. For example see Rappaport, D., "Have Intellectuals a Class Interest?" in Dreitzel, H.P., *Recent Sociology* (London: Macmillan, 1969). As John Jacobsen put it in *Technical Fouls*, p. 4, "The machine is a social product, and experts are more often 'on tap' rather than 'on top'."
8. Noble, D., *Forces of Production* (New York: Oxford University Press, 1984), p. 164.
9. *Questioning Technology*, p. 80
10. *ibid*, pp. 78–81, 88–9.
11. *ibid*, p. 95.
12. *Transforming Technology*, part III.
13. *ibid*, pp. 174–5.
14. This presumption underwrote Pickering's "dialectic of accommodations and resistances" in *The Mangle of Practice*. See *MEP* Chapter 6 for critical discussion.
15. See *MEP*, Chapter 5 for further discussion of this point.
16. See *MEP* for a detailed discussion of this point in the context of the history of experimental physics.
17. For a detailed historical discussion of the complex social processes involved in the discovery of techniques for the stabilisation, reproduction, and communication of "the electromagnetic field" in nineteenth century physics see

Gooding, D., *Experiment and the Making of Meaning* (Dordrecht: Kluwer Academic Publishers, 1990).

18. Thomas Hughes termed this a "technological momentum" and used this idea to bridge technological determinism and social constructionism by connecting both social and technical forces as causal agents for historical change. See "Technological Momentum" in Smith and Marx, eds, pp. 102–13.

19. *TT*, pp. 153–61.

20. See also Gorz, A., *Ecology As Politics* (Boston: South End Press, 1982).

21. Feenberg relied heavily upon Bruno Latour's example of an automatic door closer to elucidate this point. See Latour, B., "Where are the Missing Masses? The Sociology of a Few Mundane Artifacts" in Bijker and Law (eds), *Shaping Technology, Building Society*.

22. *Questioning Technology*, p.131.

23. See next chapter for a discussion of Barber's "strong democracy".

24. Borgmann discussed this point in detail in *Crossing the Postmodern Divide* (University of Chicago Press, 1992).

25. *Questioning Technology*, pp. 90–1

26. *TT*, p. 155.

27. *ibid*, p. 61.

Chapter 4   Participatory Democracy

1. See Plutarch's "Solon" in *The Rise and Fall of Athens* (Harmondsworth: Penguin, 1960); Thucydides' *The Peloponnesian War* (Hobbes, trans., University of Chicago Press, 1989); and Aristotle's *The Athenian Constitution* (Rhodes, trans., London: Penguin Classics, 1984). The latter was probably written by one or more students of Aristotle. See also Forrest, W.G., *The Emergence of Greek Democracy* (New York: McGraw Hill, 1966); Whitehead, D., *The Demes of Attica, 508/7 – ca. 250 BC* (Princeton University Press, 1986); and Farrar, C., "Ancient Greek Political Theory as a Response to Democracy" in Dunn, J. (ed.), *Democracy: The Unfinished Journey, 50 BC to AD 1993* (Oxford University Press, 1992).

2. Finley, M.I., *Democracy: Ancient and Modern* (London: Chatto and Windus, 1973), pp. 18–19.

3. As Robert Dahl pointed out, in *Democracy and its Critics* (New Haven: Yale University Press, 1989, p. 20), the claim that Athenian democracy should be the standard or ideal for democracy is more a product of romanticism than attendance to historical facts. David Held argued that, in *Models of Democracy* (p. 4), by modern standards, the Athenian city-state was not a democracy at all. See also Dickenson, D., *Property, Women and Politics: Subjects or Objects?* (Cambridge: Polity Press, 1997), Chapter 2.

4. Skinner, Q., "The Italian City-Republics" in Dunn, J. (ed.), *Democracy: The Unlimited Journey, 508 BC to AD 1993* (Oxford University Press, 1992); Waley, D., *The Italian City-Republics* (New York: McGraw Hill, 1969); and Mundy, J.H. and Reisenberg, P., *The Medieval Town* (New York: Van Nostrand Reinhold, 1958).

5. Barber, B., *The Death of Communal Liberty: A History of Freedom in a Swiss Mountain Canton* (Princeton University Press, 1974); Linder, W., *Swiss Democracy* (New York: St. Martin's Press, 1998).

6. Soboul, A., *The Sans-Culottes: The Popular Movement and Revolutionary Government*, 1793–1794, vol. 1 (Remy Ingles Hall, trans.), Garden City, NY: Anchor & Doubleday, 1972).

7. Rosner, M., *Participatory Political and Organizational Democracy in the Experience of the Israeli Kibbutz* (University of Haifa, 1981); Russel, R., Hanneman, R. and Getz, S., "Demographic and Environmental Influences on the Diffusion of Changes among Israeli Kibbutzim, 1990–2001" in Smith, V. (ed.), *Research in the Sociology of Work, Volume 16: Worker Participation: Current Research and Future Trends* (Amsterdam: Elsevier IAI, 2006), pp. 263–91.

8. Suny, R.G., *The Making of the Georgian Nation* (London: Routledge, 1989) and Kautsky, K., *Georgia: A Social Democratic Peasant Republic: Impressions and Observations* (London: International Bookshops Limited, 1921).

9. Bryan, F.M., *Real Democracy: The New England Town Meeting and How it Works* (University of Chicago Press, 2004); Zimmerman, J., *The New England Town Meeting* (Westport, CT: Praeger, 1999); Mansbridge, J., *Beyond Adversary Democracy* (New York: Basic Books, 1980); Cross, R.A., *The Minutemen and their World* (New York: Hill and Wang, 1976); Katz, S.N. (ed.), *Colonial America: Essays in Political and Social Development* (Boston: Little & Brown, 1971); Zuckerman, M., *Peaceable Kingdoms* (New York: Random House, 1970).

10. Roussopoulos, D., "The Case of Montreal" in Roussopoulos and Benello (eds), *Participatory Democracy*, pp. 293–324; Menegat, R., "Participatory Democracy and Sustainable Development: Integrated Urban Management in Porto Alegre, Brazil" in *Environment and Urbanization*, October 2002, pp. 1–26; Baierle, S., "The Case of Porto Alegre: The Politics and Background" and Latendresse, A., "The Case of Porto Alegre: The Participatory Budget" in Roussopoulos and Benello (eds), *Participatory Democracy*, pp. 270–86, pp. 287–91; Sonza, C., "Participatory Budgeting in Brazilian Cities: Limits and Possibilities in Build-ing Democratic Institutions" in *Environment and Urbanization*, April 2000, pp. 159–84; Chomsky, N., *Temporaes: On Democracy and Self-Management* (São Paulo: Humanitas, 1999); Wainwright, H., *Reclaim the State: Experiments in Popular Democracy* (London: Verso, 2003).

11. Valls Peirats, J., *The CNT in the Spanish Revolution, Vol. 1* (Earlham (ed.), Hastings: Meltzer Press, 2001, first published 1951), especially Chapter 15, and *The Anarchists in the Spanish Revolution* (Detroit-Toronto: Black & Red, 1976, reprinted by Freedom Press, 1990), especially Chapter 10; Earlham, C., "Revolutionary Gymnastics and the Unemployed: The Limits of the Spanish Anarchist Utopia, 1931–37" in Flett, K. and Renton, D. (eds), *The Twentieth Century: A Century of Wars and Revolutions?* (London: Rivers Oram, 2000); Bookchin, M., *To Remember Spain: The Anarchist and Syndicalist Revolution of 1936* (San Francisco: AK Press, 1994); Kelsey, G., *Anarcho-syndicalism, Libertarian Communism and the State: The CNT in Zaragoza and Aragorn* (Amsterdam: Institute for Social History, 1991); Ackelsberg, M.A. *Free Women of Spain: Anarchism and the Struggle for Emancipation of Women* (Bloomington: Indiana University Press, 1991); Guérin, D., *No Gods, No Masters* (Oakland, CA: AK Press, 2005, first published 1980), pp. 437–70; Leval, G., *Collectives in the Spanish Revolution* (London: Freedom Press, 1975); Dolgoff, S., *The Anarchist Collectives: Workers' Self-Management in the Spanish Revolution 1936–1939* (New York: Free Life Editions, 1974); Broué, P.

and Témine, E., *The Revolution and the Civil War in Spain* (University of Cambridge, 1972); Richards, V., *Lessons from the Spanish Revolution* (London: Freedom Press, 1953), pp. 46–7 and pp. 76–81.

12. On F.J. Romero Salvadó's account (in *The Spanish Civil War*, Basingstoke: Palgrave Macmillan, 2005), apart from beating up his grandfather and other criminal acts, "the anarchists" played hardly any role in the Spanish Civil War, and the CNT-FAI were "extremists" inciting "mob rule". Of course, the violent acts perpetrated by some members of the CNT-FAI (such as assassinations, bank robberies, bombings, and beating up grandfathers) should be condemned, but it is unreasonable, especially in the circumstances of a civil war, to impose a higher moral standard on the CNT-FAI members than on the Republican government forces or the Francoist forces. A civil war by its nature is a breakdown in "law and order" and many historical accounts show that the CNT-FAI showed considerable moral discipline and restraint in comparison to either the Republican or Francoist soldiers, as well as being instrumental in saving the Republic from Francoist insurrection in July 1936 by galvanising mass resistance (among its eight hundred thousand members!), arming popular militias, organising the collectives, and supplying the Republic with food and manufactured goods. In my view, Romero Salvadó was uncritically sympathetic to the Republic and personally biased against "the anarchists". As Noam Chomsky pointed out in "Objectivity and Liberal Scholarship" (first published in *American Power and the New Mandarins*, New York: New Press, 1969, pp. 23–158; reprinted in *Chomsky on Anarchism*, Oakland, CA: AK Press, 2005, pp. 11–100), this kind of bias amongst liberal academics is not new. Many of the liberal academic writings on the role of the CNT-FAI in the Civil War were overly reliant on the Republican controlled press and even on Communist Party propaganda when those sources supported their own biases and assumptions. Many of their claims about forced collectivisation or anarchists collaborating with fascists were based on unsubstantiated or false reports designed to discredit opposition to the authority of the Republican government. Chomsky was also critical of some liberal historians, such as Gabriel Jackson, for their uncritical assertion of assumptions. For example, many of Jackson's criticisms of the efficiency of the collectives in *The Spanish Republic and the Civil War: 1931–1939* (Princeton University Press, 1967) were based on the assertion that they must have been inefficient and coercive because they were not based on liberal capitalism. He did not provide any empirical evidence to support this claim. However, while it is clearly the case that Jackson was quite unsympathetic to the CNT-FAI and "the anarchists" in general, he argued that the CNT-FAI showed considerable moral restraint in dealing with their political enemies, while the Francoist forces conducted mass murders, rapes, and lootings of whole villages, tortured and slaughtered captured prisoners, executed all political opponents, and even murdered conscientious objectors among their own ranks. He also admitted that the evidence supports the CNT claim that the collectivisations were supported by the vast majority of peasants and workers. See "The Living Experience of the Spanish Civil War Collectives" (in *Newsletter of the Society for Spanish and Portuguese Historical Studies*, vol. 1, no. 2, 1970, pp. 4–11). Many of the claims regarding coercion and forced collectivisation made by Franz Borkenau in *The Spanish Cockpit* (University of

Michigan Press, 1963, first published in 1938) were based on hearsay rather than first hand experience. For example his assessment of the agrarian collective in the Aragon town of Fraga (pp. 104–12) was made after only spending one day in the town and largely based upon a conversation with one peasant in a bar! He claimed that Fraga was a village with a population of one thousand, when, in fact, it was a town with a population of over eight thousand. Many of Borkenau's assessments overly relied on Republican controlled media sources, but even Borkenau admitted that the agricultural collectives and industrial cooperatives had increased productivity, improved working conditions, and provided social provisions (p. 184). He also argued that the fall of Málaga to the fascists was a direct consequence of the refusal of the Republican government to allow the people arms to defend themselves (pp. 219–28) because it did not have popular support among workers and peasants and had to impose its authority by force (pp. 282–92). The Republican forces also attacked civilian populations, arrested, tortured, imprisoned, and executed political prisoners without trial, and even undermined the militias fighting against the Francoist forces on the front lines by seizing their supplies, redirecting Republican forces to disarm them, and using their best armed forces to suppress their political opponents behind the front lines. See Peirats' accounts of the Republican suppressions of the collectives in *The Anarchists in the Spanish Revolution*, Chapter 16, and also George Orwell's *Homage to Catalonia* (New York: Harcourt, 1952). For discussions of the role of the Stalinist Communist Party (the PSUC) in undermining the social revolution in Spain, as well as the CNT-FAI, the POUM, and the liberal democrats within the Republican government see Carr, E.H. *The Comintern and the Spanish Civil War* (London: Macmillan, 1984); Morrow, F., *Revolution and Counter-revolution in Spain* (London: New York Publications, 1963); Bolloten, B., *The Grand Camouflage: The Communist Conspiracy in the Spanish Civil War* (New York: Praeger, 1961); Brenan, G., *The Spanish Labyrinth* (Cambridge University Press, 1960, first published in 1943); and Rocker, R., *The Tragedy of Spain* (New York: Freie Arbeiter Stinnme, 1937). However, as Romero Salvadó argued (Chapter 3, pp. 152–6, and pp. 161–6) the role of the British, French, and American blockade in allowing Franco's forces to be supplied by Fascist Italy and Nazi Germany and forcing the Republic to rely on the Soviet Union for weapons and supplies should not be underestimated. See also Guttmann, A., *The Wound in the Heart: America and the Spanish Civil War* (New York: The Free Press, 1962) and Traina, R.P., *American Diplomacy and the Spanish Civil War* (Bloomington: Indiana University Press, 1968) for accounts of how the US Administration allowed American oil companies to supply Franco's forces, while bringing pressure to bear on any American company that supplied the elected and recognised Republican government, even when they were fulfilling contracts signed before the commencement of the civil war in July 1936.

13. Many original documents were destroyed during the civil war and Franco's regime systematically rewrote its history – largely attempting to remove any trace of the success of the social revolution achieved by the peasants' collectives and workers' cooperatives. See Richards, M., *A Time of Silence: Civil War and the Culture of Repression in Franco's Spain*, 1936–1945 (Cambridge University Press, 1998); Boyd, C., *Historia Patria: Politics, History, and National*

Identity in Spain, 1875–1975 (Princeton University Press, 1997), pp. 232–301; Alted, A., "Education and Political Control" in Graham, H. and Labany, J. (eds), *Spanish Cultural Studies: An Introduction* (Oxford University Press, 1995), pp. 196–201. Of course, all history is a matter of interpretation and perspective, but this makes the histories told by the people involved invaluable, even if partisan, and, in my view, the histories told by the members of the CNT-FAI, the agricultural collectives, and the industrial cooperatives are an essential part of writing history "from below", as E.P. Thomson urged historians to do, in his famous *The Making of the English Working Class* (Harmondsworth: Pelican, 1968, revised 2nd ed.), when writing histories of the Industrial Revolution in Britain. Historians, such as Peirats and Leval, being members of the CNT, obviously relied on their own perspective and interpretations in their editorial selections and judgements, and it is important to be aware of this, as both Peirats and Leval implored their readers to do, but it is somewhat disingenuous for academics, such as Hugh Thomas, in his essay "Anarchist Agrarian Collectives in the Spanish Civil War", in Gilbert, M. (ed.), *A Century of Conflict, 1850–1950: Essays for A.J.P. Taylor* (New York: Atheneum Publishers, 1967, pp. 245–63), to rely on Peirats' and Leval's 'first hand' accounts (as his primary sources!) while simultaneously dismissing them for being selective and biased.

14. *The Social Contract* Book III, Chapter 4: "If there were a nation of gods, it would govern itself democratically. A government so perfect is not suited to men." (London: Penguin, 1968, p. 114).

15. Benello, C.G., *From the Ground Up: Essays on Grassroots and Workplace Democracy* (Boston, Mass: South End Press, 1992).

16. Roussopoulos, D. and Benello, C.G. (eds), *Participatory Democracy: Prospects for Democratising Democracy* (New York: Black Rose Books, 2005).

17. Rocker, R., *Anarcho-Syndicalism: Theory and Practice* (Oakland, CA: AK Press, 2004), first published in 1938.

18. Pannekoek, A., *Workers' Councils* (Edinburgh: AK Press, 2003), first published in 1950; Guérin, D., "In Search of a New Society" in *Anarchism* (New York: Monthly Review Press, 1970) pp. 41–69; Bookchin, M., "Towards a Liberatory Technology" in Roussopoulos and Benello (eds), *Participatory Democracy*, pp. 85–126; Chomsky, N., "The Relevance of Anarcho-syndicalism" BBC *London Weekend TV* interview with Peter Jay, July 25, 1976, transcript published in *Radical Priorities* (Otero, ed., Oakland, CA: AK Press, 2003), pp. 211–24, and, in *Chomsky On Anarchism* (Oakland, CA: AK Press, 2005), pp. 133–48. See also Errico Malatesta's article "Post-Revolutionary Property System", first published in 1929 and reprinted in *The Anarchist Revolution: Polemical Articles 1924–1931* (London: Freedom Press, 1995), pp. 113–19.

19. Godwin, W., *Enquiry Concerning Political Justice and its Influence on General Virtue and Happiness* (State University of New York Press, 1972).

20. Cole, G.D.H., *Self-Government in Industry* (London: Bell & Sons, 1919) and *Guild Socialism Restated* (London: Leonard Parsons, 1920). Bertrand Russell's *Roads to Freedom: Socialism, Anarchism, and Syndicalism* (London: Routledge, 1970), first published in 1918, drew heavily on the Cole's ideas.

21. Pateman, C., *Participation and Democratic Theory* (Cambridge University Press, 1970) and *The Problem of Political Obligation: A Critique of Liberal Theory* (Cambridge: Polity Press, 1985); Macpherson, C.B., *The Real World of Democracy*

(Oxford University Press, 1966) and *The Life and Times of Liberal Democracy* (Oxford University Press, 1977).
22. Barber, B., *Strong Democracy: Participatory Politics for a New Age* (Berkeley: University of California Press, 1984).
23. See also Macpherson, C.B., *The Life and Times of Liberal Democracy* (Oxford University Press, 1977); Bobbio, N., *Democracy and Dictatorship* (Cambridge: Polity Press, 1989); Held, D., *Models of Democracy* (Stanford University Press, 1987) and *Democracy and the Global Order: From the State to Cosmopolitan Governance* (Cambridge: Polity Press, 1995).
24. Kanter, R.M., *Commitment and Community* (Cambridge, MA: Harvard University Press, 1972); Nisbet, R.A., *The Quest for Community* (New York: Oxford University Press, 1976); Oldfield, A., *Citizenship and Community: Civic Republicanism and the Modern World* (London: Routledge, 1990); Putnam, R.D., *Making Democracy Work: Civic Tradition in Modern Italy* (Princeton University Press, 1993); Oliver, D. and Heater, D., *The Foundations of Citizenship* (Hemel Hempstead: Harvester Wheatsheaf, 1994); Dagger, R., *Civic Virtues: Rights, Citizenship and Republican Liberalism* (Oxford University Press, 1997); Tam, H., *Communitarianism: A New Agenda for Politics and Citizenship* (Basingstoke: Macmillan, 1998).
25. See also Demaine, J. and Entwistle, H. (eds), *Beyond Communitarianism: Citizenship, Politics and Education* (Basingstoke: Palgrave Macmillan, 1996).
26. *Strong Democracy*, p. 118.
27. See also Ellul, J., "Technology and Democracy" in *Democracy in a Technological Society*, pp. 35–50.
28. Biehl, J., *The Politics of Social Ecology: Libertarian Municipalism* (New York: Black Rose, 1999); Schuman, M.H., *Going Local: Creating Self-Reliant Communities in a Global Age* (New York: Free Press, 1998); Bookchin, M., *From Urbanization to Cities* (London: Cassell, 1995); Berry, J.M., Portney, K.E. and Thomson, K., *The Rebirth of Urban Democracy* (Washington, DC: Brookings Institution, 1993); Castells, M., *The City and the Grassroots: A Cross-Cultural Theory of Urban Social Movements* (Berkeley: University of California Press, 1983); Roussopoulos, D. (ed.), *The City and Radical Social Change* (Montreal: Black Rose, 1982); Schecter, S., *The Politics of Urban Liberation* (Montreal: Black Rose, 1978); Mumford, L., *The City in History* (New York: Harcourt, Brace and World, 1961).
29. See also Suttles, G., *Social Order of the Slum* (University of Chicago Press, 1968); Kotler, M., *Neighborhood Government* (New York: Bobbs-Merrill, 1969); Yates, D., *Neighborhood Democracy* (Lexington, MA: D.C. Heath, 1973); and Crenson, M.A., *Neighborhood Politics* (Cambridge, MA: Harvard University Press, 1983).
30. J.S. Mill also considered this to be essential for the health of representative government.
31. Heater, D., *Citizenship* (London: Longman, 1999).
32. Tocqueville, A. wrote in *Democracy in America* (2 vols., London: Fontana, 1968) (p. 76) that town meetings "are to liberty what primary schools are to science; they bring it within the people's reach, they teach men how to use and enjoy it". See also Bryan, *Real Democracy* (pp. 286–92) for a discussion of the educational value of town meetings for the development of civic education.
33. See Gould, C., *Rethinking Democracy* (Cambridge University Press, 1988), and Philips, A., *Democracy and Difference* (Cambridge: Polity Press, 1993) for detailed discussions of this point.

34. For example, see Frazer, E., and Lacey, N., *The Politics of Community* (Hemel Hempstead: Harvester Wheatsheaf, 1993).
35. Arendt, H. *The Human Condition* (University of Chicago Press, 1958), p. 59.
36. See also Pateman, *Participation & Democratic Theory*, Chapter 3.
37. See also Sale, K., *Human Scale* (New York: Coward McCann and Geoghegan, 1980) and Follet, M.P. *The New State: Group Organization, the Solution of Popular Government* (Magnolia, MA: Peter Smith, 1965, first published in 1919).
38. Corcoran, P., "The Limits of Democratic Theory" in Duncan, G. (ed.), *Democratic Theory and Practice* (Cambridge University Press, 1983), pp. 13–24.
39. Robert Dahl first proposed the polyarchic theory of democracy (rule by competing minorities, institutions, and organisations) in *Preface to Democratic Theory* (University of Chicago Press, 1956) and further developed it in *Polyarchy: Participation and Opposition* (New Haven, CT: Yale University Press, 1971). See also Sartori, G., *Democratic Theory* (Detroit: Wayne State University Press, 1962); Sharp, G., *Social Power and Political Freedom* (Boston: Porter Sargent, 1980) for discussions of how competing and powerful groups can confront organised power through collective action and alliances. For a discussion of polyarchic interactions between powerful industries, state departments, NGOs, activist movements, political parties, corporations, pressure groups, and unions, and how these influenced and effected changes in the legislation, implementation, and enforcement of environmental and public health regulations, see Clegg, H., *A New Approach to Industrial Democracy* (Oxford: Blackwell, 1960); Eckstein, H., *Division and Cohesion in Democracy* (Princeton University Press, 1966); and Beck, U., *The Risk Society* (London: Sage, 1992).
40. *Preface to Democratic Theory*, p. 132.
41. See Truman, D.B., *The Governmental Process* (New York: Alfred Knopf, 1951) for an interesting account of the importance of over-lapping membership for democratic participation. Truman described government as "a protean complex of criss-crossing relationships that change its strength and direction with alterations of power and the standing of interests". (p. 508)
42. Rawls, J., "The Idea of Public Reason Revisited" in *University of Chicago Law Review*, 64, 3, 1997, pp. 765–6.
43. Galston, *Liberal Pluralism*, pp. 40–9, also made this criticism of Rawl's position.
44. *ibid*, pp. 66–9.
45. *Democracy and its Critics*, pp. 182–3.
46. *P&DT*, Chapter 2.
47. In *Emile*, Book IV (London: Penguin Classics, 1968), Rousseau argued that it is our physical weaknesses and vulnerabilities that necessitate human sociability, reasoned deliberation, and cooperation.
48. *Social Contract*, Book I, Chapter 7. It is for this reason that Rousseau considered the imprisonment of criminals to have the dual aspects of protecting society and educating the errant individual, allowing them to exercise reason and free themselves from their passions.
49. As Pateman argued, in *The Problem of Political Obligation* (pp. 157–8), Rousseau's exclusion of women from the citizenry was contingent upon his belief that women were prone to "immoderate passions" that lead to a lesser capacity to overcome the enslavement to the passions through the exercise

of reason. Poor men would be excluded if they were dependent upon other men because they did not own their own property and livelihood.

50. Dahl, R.A. and Tufte, E.R., *Size and Democracy* (Stanford University Press, 1973) and Bookchin, M., *The Limits of the City* (New York: Harper and Row, 1974). See also Mansbridge, *Beyond Adversary Democracy*, pp. 59–76. However, Bryant, in his thirty year study of over 1500 New England town meetings, has not found that there is any correlation between the proportion of "newcomers" to "natives" and levels of participation see *Real Democracy*, p. 122, and socio-economic diversity has little effect on either levels or equality of participation, pp. 185–6. Dahl also noted this about New England town meetings in *On Democracy* (New Haven: Yale University, 1998), p. 111.
51. *Human Condition*, p. 176.
52. Arendt, H., *On Revolution* (London: Penguin Classics, 2006, first published 1963), pp. 32–3.
53. *Op cit*, p. 57.
54. *ibid*, pp. 178–80.
55. *SD*, pp. 164–7.
56. *ibid*, p. 170.
57. *ibid*, p. 174.
58. *On Revolution*, p. 136.
59. *ibid*, p. 166.
60. For examples see Dijas, M., *The New Class* (New York: Praeger, 1957); Pateman, *Participation and Democratic Theory* Chapter 5; Hunnius, G., "The Yugoslav System of Decentralisation and Self-Management" in Roussopoulos and Benello (eds), *Participatory Democracy*, pp. 127–56. For an official Yugoslav account of the organisation of the Communes see Milivojevic, D., *The Yugoslav Commune* (Belgrade: Medunarodna Politika, 1965). For an official Yugoslav model of the "self-management system" see Vlajic, V., *The Working Units* (Belgrade: The Central Council of the Confederation of Trade Unions of Yugoslavia, 1967).
61. For an interesting and detailed critique of the Yugoslav "self-management system" see Lydall, H., *Yugoslav Socialism Theory & Practice* (Oxford: Clarendon Press, 1984). For general theoretical discussions see Vanek, J., *The General Theory of Labour-Managed Market Economies* (Cornell University Press, 1970); Neuberger, E. and James, E., "The Yugoslav Self-Management System" in Borenstein, M. (ed.), *Plan and Market* (Yale University Press, 1973); and Granick, D., *Enterprise Guidance in Eastern Europe* (Princeton University Press, 1975).
62. Thomas, H. and Logan, C., *Mondragón: An Economic Analysis* (London: Allen & Unwin, 1982); Whyte, W.F. and White, K.K., *Making Mondragón: The Growth and Dynamics of the Worker Cooperative Complex* (Ithaca: ILR Press, 1991); Kasmir, S., *The Myth of Mondragón: Cooperatives, Politics, and Working Class Life in a Basque Town* (Albany, NY: State University of New York Press, 1996); Macleod, G., *From Mondragón to America: Experiments in Community Economic Development* (Sidney, Nova Scotia: University College of Cape Breton Press, 1997).
63. Coates, K. (ed.), *The New Worker Cooperatives* (New York: Spokesman Books, 1976); Rothschild, J. and Whitt, J.A., *The Cooperative Workplace: Potentials and Dilemmas of Organizational Democracy and Participation* (Cambridge University Press, 1986); Cornforth, C., Thomas, A., Lewis, J. and Spear, R., *Developing Successful Worker Cooperatives* (London: Sage, 1988); Pencavel, J., *Worker*

Participation: Lessons from the Worker Co-ops of the Pacific Northwest (New York: Russell Sage Foundation, 2001); Smith, V. (ed.), *Research in the Sociology of Work, Volume 16: Worker Participation: Current Research and Future Trends* (Amsterdam: Elsevier IAI, 2006).

64. Moody, K., *Workers in a Lean World* (London: Verso, 1997) and Moore, T.S., *The Disposable Workforce* (New York: De Gruyter, 1996).

65. Melman, S., *Decision-Making and Productivity* (London: Blackwell, 1958); Blauner, R., *Alienation and Freedom: The Factory Worker and his Industry* (University of Chicago Press, 1966); Jacques, E., *Employee Participation and Managerial Authority* (London: Tavistock, 1968); Noble, D.F., *Forces of Production A Social History of Automation* (New York: Oxford University Press, 1985). See also Kunda, G., *Engineering Culture: Control and Commitment in a High-Tech Culture* (Philadelphia: Temple University Press, 1993).

66. For interesting discussions of workers' control, industrial cooperatives, and democratic participation in workplaces and economic development see Coates, K. and Topham, T. (eds), *Industrial Democracy in Great Britain* (London: MacGibbon and Kee, 1967); Blumberg, P., *Industrial Democracy: The Sociology of Participation* (London: Constable, 1968); Vanek, J., *The Labour-Managed Economy* (Ithaca, NY: Cornell University Press, 1977); Zwerdling, D., *Workplace Democracy* (New York: Harper & Row, 1978); Gunn, C.E., *Workers' Self-Management in the United States* (Ithaca, NY: Cornell University Press, 1984); Greenberg, E.S., *Workplace Democracy: The Political Effects of Participation* (Ithaca: Cornell University Press, 1986); Devine, P., *Democracy and Economic Planning* (Cambridge: Polity Press, 1988); Adams, F.T. and Hanson, G.B., *Putting Democracy to Work* (San Francisco: Berret-Koehler, 1992); Ellerman, D., *Property and Contract in Economics: The Case for Economic Democracy* (Cambridge, MA: Blackwell, 1992); Krimerman, L. and Lidenfeld, F. (eds), *When Workers Decide: Workplace Democracy Takes Root in North America* (Philadelphia: New Society Publisher, 1992); Lichtenstein, N. and Harris, H.J. (eds), *Industrial Democracy in America* (Cambridge University Press, 1993); Harris, Z., *The Transformation of Capitalist Society* (Lanham, MD: Rowan & Littlefield, 1997); Markey, R. (ed.), *Innovation and Employee Participation through Workers' Councils: International Case Studies* (Brookfield: Ashgate, 1997); Melman, S., *After Capitalism: From Managerialism to Workplace Democracy* (New York: Knopf, 2001); Albert, M., *Parecon: Life after Capitalism* (London: Verso, 2003); Gillespie, J., "Towards Freedom in Work" in Roussopoulos and Benello (eds), *Participatory Democracy*, pp. 64–84.

67. Vanek, J., *The Participatory Economy* (Ithaca, NY: Cornell University Press, 1971); Lidenfeld, F., and Rothschild-Whitt, J., *Workplace Democracy and Social Change* (Boston: Porter Sargent Publishers, 1982); Dahl, *A Preface to Economic Democracy*, and, Benello, *From the Ground Up*, Chapter 10.

68. Vanek, *The Participatory Economy*.

69. Benello, C.G., "Group Organization and Socio-Political Structure" in Roussopoulos and Benello (eds), *Participatory Democracy*, pp. 34–49.

70. *Yugoslav Socialism: Theory & Practice*, p. 210.

71. In many respects, the Communist Party in Cuba relied on this even more than the Bolsheviks in the early stages of the industrialisation of the Soviet Union. See Bengelsdorf, C., *The Problem of Democracy in Cuba: Between Vision and Reality* (Oxford University Press, 1994).

72. Sclove, R.E., *Democracy and Technology* (New York: Guildford Press, 1995), p. 114. See also Chisholm, D., *Coordination without Hierarchy: Informal Structures in Multi-organisational Systems* (University of California Press, 1989), for further examples and discussion.

Chapter 5   Science and Technology

1. Dunn, J., *Western Political Theory in the Face of the Future* (Cambridge University Press, 1979).
2. Barney, D., *Prometheus Wired: The Hope for Democracy in the Age of Network Technology* (University of Chicago Press, 2000), p. 25.
3. Fishkin, J., *Democracy and Deliberation* (New Haven, CT: Yale University Press, 1991).
4. *MSCN*, Chapters 5 and 6.
5. Kleinman, "Democratizations of Science and Technology" in Kleinman (ed.), *Science, Technology & Democracy*, pp. 139–65.
6. Sclove, R.E., *Democracy and Technology* (London: Guildford Press, 1995), p. 5.
7. *ibid*, Chapter 3.
8. Sclove attributed the coinage of this term to Anthony Giddens, *Central Problems in Social Theory: Action, Structure, and Contradiction in Social Analysis* (University of California Press, 1979), Chapter 2.
9. Polanyi, M., *Science, Faith, and Society: A Searching Examination of the Meaning and Nature of Scientific Inquiry* (University of Chicago Press, 1964). First published in 1946, this book contains Polanyi's counter to Marxist theories of science developed in the Soviet Union during the assertion of Lysenko's theories of genetics and biology. Polanyi's argument was a defence of the independence of scientific research from social planning, understood in the context of events in the Soviet Union, while he acknowledged that, at bottom, this defence involved little more than a partisan defence of Western science against Soviet science, appealing to faith (or trust) in the former and scepticism (or distrust) in the latter. For interesting and detailed studies of Lysenko and the suppression of geneticists in the Soviet Union see Roll-Hansen, N., *The Lysenko Effect: The Politics of Science* (New York: Humanity Books, 2005) and Joravsky, D., *The Lysenko Affair* (University of Chicago Press, 1970). It is important to note that the Lysenko case does not reveal inherent dangers to democratic participation in science, as some ill informed commentators have asserted, but, rather, shows the dangers inherent to centralised control over science in a highly bureaucratic and authoritarian society.
10. In this respect, Polanyi's philosophy of science pre-empted Roy Bhaskar's *A Realist Theory of Science* (Leeds Books, 1975). For critical discussion of Bhaskar's realism see *MEP*, Chapter 2.
11. *SFS*, p. 25.
12. See *MEP*, Chapters 2 and 3.
13. For detailed discussion, *ibid*, Chapter 6.
14. *Op cit*, p. 40.
15. Popper, K.R., *The Logic of Scientific Discovery* (London: Hutchinson, 1959).
16. *Op cit*, p. 45.

17. *ibid*, pp. 31–4.
18. *ibid*, p. 38.
19. Kuhn, T.S., *The Structure of Scientific Revolutions* (University of Chicago Press, 1962).
20. *Op cit*, p. 48.
21. Polanyi pre-empted sociological accounts of science, such as David Bloor's *Knowledge and Social Imagery* (Oxford: Routledge, 1976); Bruno Latour & Steve Woolgar's *Laboratory Life: The Social Construction of Scientific Facts* (Beverly Hills, Cal: Sage, 1979); Harry Collin's *Changing Order: Replication and Induction in Scientific Practice* (London: Sage, 1985); Karin Knorr-Cetina's *The Manufacture of Knowledge* (Pergamon Press, 1981), and Steve Shapin's *A Social History of Science* (University of Chicago Press, 1994).
22. *Op cit*, p. 73.
23. *ibid*, pp. 16–17.
24. *ibid*, p. 57.
25. *ibid*, pp. 64–5.
26. *ibid*, p. 60.
27. *ibid*, p. 61.
28. *ibid*, p. 67.
29. *ibid*, p. 68.
30. Bakunin, M., *God and the State* (New York: Dover, 1970), pp. 30–1, first published in 1882.
31. *ibid*, p. 32.
32. *ibid*, pp. 62–4.
33. Drahas, R. and Mayne, R. (eds), *Global Intellectual Property Rights: Knowledge, Access, and Development* (Houndmills: Palgrave Macmillan, 2002).
34. Rosenberg, N., *Inside the Black Box: Technology and Economics* (Cambridge, Mass: MIT Press, 1982); Flamin, K., *Targeting the Computer* (Washington, DC.: Brookings Institution, 1987); Smith, M.R., *Military Enterprise and Industrial Technology* (Cambridge, Mass: MIT Press, 1987); Henderson, J., *The Globalization of High Technology* (London: Routledge, 1989); Lockeretz, W. and Anderson, M.D., *Agricultural Research Alternatives* (Lincoln: University of Nebraska Press, 1993); Bonanno, A., Busch, L., Friedland, W., Gouveia, L. and Mingione, E. (eds), *From Columbus to ConAgra: The Globalization of Agriculture and Food* (Lawrence: University Press of Kansas, 1994); Forman, P. and Sanchez-Ron, J.M. (eds), *National Military Establishments and the Advancement of Science and Technology: Studies in 20ᵗʰ Century Science* (Dordrecht: Kluwer Academic, 1996).
35. Snow, C.P., *Two Cultures and a Second Look* (Cambridge University Press, 1964).
36. For examples see Rheingold, H., *Virtual Reality* (New York: Summit Books, 1991); Negroponte, N., *Being Digital* (New York: Knopf, 1995); Grossman, L.K., *The Electronic Republic: Reshaping Democracy in the Information Age* (New York: Viking Press, 1995); Edwards, P., *The Closed World: Computers and the Politics of Discourse in Cold War America* (Cambridge, Mass.: MIT Press, 1996); Tsagarousianou, R., Tambini, D. and Bryan, C. (eds), *Cyberdemocracy: Technology, Cities and Civic Networks* (London: Routledge, 1998).
37. For examples see Weizenbaum, J., *Computer Power and Human Reason: From Judgement to Calculation* (San Francisco: Freeman, 1976); Lyon, D., *The*

Information Society: Issues and Illusions (Cambridge, MA: Polity, 1988); Salvaggio, J.L. (ed.), *The Information Society: Economic, Social and Structural Issues* (Hillsdale, NJ: Lawrence Erlbaum, 1989); Traber, M., *The Myth of the Information Revolution: Social and Ethical Implications of Communication Technology* (London: Sage, 1986).

38. Barney, D., *Prometheus Wired: The Hope for Democracy in the Age of Network Technology* (University of Chicago Press, 2000).

39. *ibid*, p. 118.

40. *ibid*, p. 190.

41. *Technology and the Character of Contemporary Life*, pp. 40–8

42. *Crossing the Postmodern Divide* (University of Chicago Press, 1992).

43. See Kellner, D., "Crossing the Postmodern Divide with Borgmann, or Adventures in Cyberspace" in *Technology and the Good Life?*, pp. 234–55, for further discussion of this point in the context of computers and network technologies. Kellner also argued that Borgmann's philosophy of technology entails an ontological realism, underwriting his ideas of authenticity and engagement, which is at odds with the perspectivism and constructionism inherent to postmodern critiques. It is this ontological realism upon which Borgmann's distinction between the hyper-realism of device paradigm and the focal realism of focal practices and things greatly depends. Kellner argued that this distinction tends to betray Borgmann's "technophobic refusal 'to see' how technology can enhance our lives, can be put to progressive political uses... that counters destructive and life-negating technologies and applications". (p. 243)

44. Feenberg, *Transforming Technology*, pp. 91–130; Barber, *Strong Democracy*, p. xv.

45. See Schneider, S., "Is the Citizen-Scientist an Oxymoron?" in Kleinman (ed.), *Science, Technology & Democracy*, pp. 103–20.

46. For a discussion of the Danish "consensus conference" experiment and its possible development in the USA see Sclove, R.E., "Town Meetings on Technology: Consensus Conferences as Democratic Participation" in Kleinman (ed.), *Science, Technology & Democracy*, pp. 33–48; for discussions of the idea of "voter juries" or "citizens' juries" as a means of public participation in the deliberation of public policy see Fishkin, J., *Democracy and Deliberation* (New Haven, CT: Yale University Press, 1991) and also Burnheim, J., *Is Democracy Possible? The Alternative to Electoral Politics* (Berkeley: University of California Press, 1985).

47. Canadian "freenets" are an example of CCCs used to aid civic communication, cooperation, and rejuvenation, see Bourque, P. and, Dickson, R., *Freenet: Canadian Online Access the Free and Easy Way* (Toronto: Stoddard, 1996).

48. Levitt, N. and Gross, P., "The Perils of Democratizing Science" in *The Chronicle of Higher Education*, October 5th 1994.

49. For examples see Petersen, J.C. (ed.), *Citizen Participation in Science Policy* (Amherst: University of Massachusetts Press, 1984); Dickson, D., *The New Politics of Science* (Chicago University Press, 1988); Krimsky, S. and Plough, A., *Environmental Hazards: Communicating Risks as a Social Process* (Dover, MA.: Auburn House, 1988); Balogh, B., *Chain Reaction: Expert Debate and Public Participation in American Commercial Nuclear Power, 1945–1975* (Cambridge University Press, 1991); Renn, O., Webler, T. and Weidemann, P. (eds), *Fairness*

and Competence in Citizens Participation: Evaluating Models for Environmental Discourse (Dordrecht: Kluwer Academic, 1995); Kleinman, D.L. (ed.), *Science, Technology & Democracy* (New York: SUNY, 2000).

50. For examples see Dunlap, T., *DDT: Scientists, Citizens, and Public Policy* (Princeton University Press, 1981); Petersen, J. (ed.), *Citizen Participation in Science Policy* (Amherst: University of Massachusetts Press, 1984); and Epstein, S., *Impure Science: AIDS, Activism, and the Politics of Knowledge* (Berkeley: University of California Press, 1996).

51. Andrews, L., *The Clone Age: Adventures in the New World of Reproductive Technology* (New York: Henry Holt, 1999). Andrews (the former chair of Human Genome Projects' working group on ethical, legal, and social implications) claimed that out of the three billion dollars awarded to the HGP only twenty five thousand dollars was awarded to the working group.

52. *Liberal Pluralism*, p. 85.

53. "Misunderstood Misunderstandings: Social Identities and Public Uptake of Science" in Irwin, A. and Wynne, B. (eds), *Misunderstanding Science? The Public Reconstruction of Science and Technology* (Cambridge University Press, 1996, pp. 19–49), pp. 43–4. See also Wynne, B., "May the Sheep Safely Graze?: A Reflexive Review of the Expert-Local Knowledge Division" in Lash, S., Szerszynski, B. and Wynne, B. (eds), *Risk, Environment and Modernity: Towards a New Ecology* (London: Sage, 1996).

54. As Jacobsen argued in *Technical Fouls*, p. 23, it is unreasonable to expect scientists and technicians to be any more courageous, insightful, or ethical than any other person, and, like most people, scientists also tend to focus on matters which are of concern or interest to them.

55. For further discussion of this point, see Harrison, P., *The Greening of Africa: Breaking Through the Battle for Land and Food* (London: Paladin Grafton Books, 1987); Appfel-Marglin, F. and Marglin, S. (eds), *Decolonising Knowledge: From Development to Dialogue* (Oxford: Clarendon Press, 1996).

56. Benello, C.G., "Group Organization and Socio-Political Structure" in Roussopoulos and Benello (eds), *Participatory Democracy*, pp. 34–49, p. 42.

57. For case studies of the problems faced by scientists and engineers when making judgements in the face of uncertainty and ambiguity, see Kahneman, D., Slovic, P. and Tversky, A., *Judgement under Uncertainty* (New York: Cambridge University Press, 1982); Krimsky, S. and Plough, A., *Environmental Hazards: Communicating Risks as a Social Process* (Dover, MA.: Auburn House, 1988); and, Beck, U., *The Risk Society* (London: Sage, 1992).

Chapter 6   Towards a Rational Society

1. *Towards A Rational Society: Student Protest, Science, and Politics* (Shapiro, trans., Boston: Beacon Press, 1970).

2. *Communication and the Evolution of Society* (McCarthy, trans., Boston: Beacon Press, 1979).

3. *The Theory of Communicative Action* (2 volumes, McCarthy, trans., Cambridge: Polity Press, 1990, 1992).

4. "Technology and Science as "Ideology"", published in *Towards a Rational Society*.

5. *Theory of Communicative Action*, vol. 2, p. 303.
6. *MSCN.*
7. *MEP.*
8. *Communication and the Evolution of Society*, p. 140.
9. *Knowledge and Human Interests*, p. 194.
10. Arendt, *On Revolution*, p. 173.
11. *Liberal Pluralism*, p. 70.
12. Havel, V., *The Power of the Powerless* (New York: Sharpe, 1979).
13. Arguably, Western countries have not quite reached the level of "post-totalitarianism" when they use state-terrorist techniques and erode the civil rights of their citizens (such as *habeas corpus*), arresting, imprisoning, torturing, and even murdering citizens that are considered to be a threat to "our way of life", labeling them as "suspected terrorists" or "enemy combatants" without due legal process or any respect for human rights. In this regard, the difference between Stalinist regimes and Western countries is merely one of technical sophistication in the means of domination.
14. *The Constitution of the United States of America* (Bedford, Mass: Applewood Books, n.d.).
15. The definition of "unreasonable searches and seizure" as being those performed "without a warrant" has been consistently upheld by the US Supreme Court, with only the limited exceptions defined as being those comprehensively and exclusively authorised by Congress under the provisions of the 1978 Foreign Intelligence Surveillance Act, such as during the first 15 days after the declaration of war and within a 72 hour period prior to obtaining a retrospective warrant from a FISA court judge, providing that there is the reasonable expectation of obtaining a FISA warrant. While the provisions of both the Constitution and FISA only protect citizens of the USA, these laws strictly prohibit the authorisation of warrantless surveillance if there is a reasonable expectation that any US citizen could be subjected to the surveillance.
16. As the Center for Constitutional Rights argued in their recent publication *Articles of Impeachment Against George W. Bush* (New York: Melville House, 2006, p. 34) "a data-mining program where the government scanned the contents of every international email and phone call for individual words or patterns of activity would be so inherently incompatible with such safeguards that it would be impossible to run if any such requirements were applied to it".
17. Mouffe, C. (ed.), *Dimensions of Radical Democracy* (London: Verso, 1992).
18. Lister, R., *Citizenship: Feminist Perspectives* (Basingstoke: Palgrave Macmillan, 2003). See also Siim, B., *Gender and Citizenship* (Cambridge University Press, 2000); Pateman, C., *The Disorder of Women* (Cambridge: Polity Press, 1989) and *The Sexual Contract* (Cambridge: Polity Press, 1988).
19. Vandenberg, A. (ed.), *Citizenship and Democracy in a Global Era* (Basingstoke: Macmillan, 2000).
20. *Citizenship*, p. 5. See also Marshall, T.H. *Citizenship and Social Class* (Cambridge University Press, 1950) and Stevenson, N. (ed.), *Culture and Citizenship* (London: Sage, 2001).
21. See Hutton, W., *The State We're In* (London: Jonathan Cape, 1995) for how the notion of citizenship has emerged in the United Kingdom from negotiations

between the socialist ideal of the State (based on equality and social justice) and the liberal tradition of individual rights and responsibilities. See also Plant, R., *Citizenship, Rights and Socialism* (London: Fabians Society, 1988).

22. *Citizenship*, p. 39.
23. See also Sassower, R., *Confronting Disaster: An Existential Approach to Techno-science* (Oxford: Lexington Books, 2004).
24. Villa, D., *Socratic Citizenship* (Princeton University Press, 2000); see also Patočka, J., *Heretical Essays in the Philosophy of History* (Dodd (ed.), Kohak, trans., Chicago: Open Court, 1996) for Socratic discussions of how freedom is historically developed through an ongoing, critical reflection upon the meaning of human existence, the metaphysical foundations of morality and epistemology. This is not to say that we should accept whatever par-ticular metaphysical or transcendental foundations seem true to us, but, rather, it is the general pursuit of such foundations and critical reflection upon prior efforts that provides our truths with any meaning at all. The possibility of freedom depends on a philosophical engagement with the mean-ing of freedom for others within an historically developed society, and, as a concept it existentially emerges on an intellectual plane between historical, philosophical, and political understandings of the meaning of being human.
25. Lee, K.N., *Compass and Gyroscope: Integrating Science and Politics for the Environment* (Washington, D.C.: Island Press, 1993); Graedel, T.E. and Allenby, B.R., *Industrial Ecology* (New York: Prentice Hall, 1995).
26. Roussopoulos and Benello (eds), *Participatory Democracy*, p. 3.
27. *Liberal Pluralism*, p. 123.
28. Kropotkin, P., *Anarchism, its Philosophy and Ideal* (London: Freedom Press, 1895), p. 43.
29. *Technics and Civilization* (New York: Harcourt Brace & Company, 1962), first published 1934.
30. *Participatory Democracy*, p. 6.
31. For further discussion see Breines, W., *Community and Organization in the New Left, 1962–1968: The Great Refusal* (New Brunswick, NJ: Rutgers Univer-sity Press, 1989) and Polletta, F., *Freedom is an Endless Meeting: Democracy in American Social Movements* (University of Chicago Press, 2002).
32. *Participatory Democracy*, p. 9.
33. *Revolt of the Masses*, pp. 75–6.
34. See Callinicos, A., *The Revenge of History: Marxism and the East European Revo-lutions* (Cambridge: Polity Press, 1991).
35. Fukuyama, F., *The End of History and the Last Man* (London: Hamish Hamilton, 1992).

Index